LLA
(CM)

543.00724
MOR

10186117

Chemometrics:
Experimental Design

D1584501

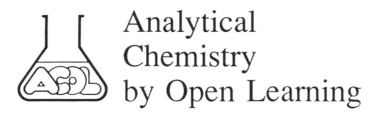

Analytical Chemistry by Open Learning

Project Director
BRIAN R CURRELL
Thames Polytechnic

Project Manager
JOHN W JAMES
Consultant

Project Advisors
ANTHONY D ASHMORE
Royal Society of Chemistry

DAVE W PARK
Consultant

Administrative Editor
NORMA CHADWICK
Thames Polytechnic

Titles in Series:

Chemometrics: Experimental Design

Analytical Chemistry by Open Learning

ED MORGAN
The Polytechnic of Wales

Published on behalf of ACOL, London
by
JOHN WILEY & SONS
Chichester · New York · Brisbane · Toronto · Singapore

Copyright © 1991 Thames Polytechnic, London, UK
Published in 1991 by John Wiley & Sons Ltd.
               Baffins Lane, Chichester
               West Sussex PO19 1UD, England
               National Chichester (01243) 779777
               International +44 1243 779777

First published as a paperback in 1995

Reprinted November 1995

All rights reserved.

No part of this book may be reproduced by any means,
or transmitted, or translated into a machine language
without the written permission of the publisher.

Other Wiley Editorial Offices

John Wiley & Sons, Inc., 605 Third Avenue,
New York, NY 10158-0012, USA

Jacaranda Wiley & Sons Ltd, 33 Park Road, Milton,
Queensland 4064, Australia

John Wiley & Sons (Canada) Ltd, 22 Worcester Road,
Rexdale, Ontario M9W 1L1, Canada

John Wiley & Sons (SEA) Pte Ltd, 37 Jalan Pemimpin #05-04,
Block B, Union Industrial Building, Singapore 2057

British Library Cataloguing in Publication Data

A catalogue record for this book is available from the British Library

ISBN 0 471 95832 8

Printed and bound in Great Britain by Redwood Books, Trowbridge, Wilts.
This book is printed on acid-free paper responsibly manufactured from
sustainable forestation, for which at least two trees are planted for each
one used for paper production.

The ACOL Series

This series of easy to read, user-friendly texts has been written by some of the foremost lecturers in Analytical Chemistry in the United Kingdom. The texts are designed for training, continuing education and updating of all technical staff concerned with Analytical Chemistry.

These texts are for those interested in Analytical Chemistry and instrumental techniques who wish to study in a more flexible way than traditional institute attendance or to augment such attendance.

Analytical Chemistry by Open Learning, ACOL, provides training resources—books, courses and computer programs. Courses based on ACOL, incorporating practical workshops and tutor assessment, are provided by Thames Polytechnic in conjunction with the Royal Society of Chemistry. BTEC qualifications are awarded for successful completion of courses and lead to a diploma that would be an appropriate preparation for applicants to the Royal Society of Chemistry for Licentiateship status, LRSC.

Thames Polytechnic will continue to support these 'Open Learning Texts', to continually refresh and update the material and to extend their coverage. For further information on ACOL, the RSC courses and the BTEC diploma, please contact:

The ACOL Office,
Thames Polytechnic,
Avery Hill Road,
Eltham,
London
SE9 2HB,
U.K.

How to Use an Open Learning Text

Open learning texts are designed as a convenient and flexible way of studying for people who, for a variety of reasons, cannot use conventional education courses. You will learn from this text the principles of one subject in Analytical Chemistry, but only by putting this knowledge into practice, under professional supervision, will you gain a full understanding of the analytical techniques described.

To achieve the full benefit from an open learning text you need to plan your place and time of study.

- Find the most suitable place to study where you can work without disturbance.

- If you have a tutor supervising your study discuss with him, or her, the date by which you should have completed this text.

- Some people study perfectly well in irregular bursts, however most students find that setting aside a certain number of hours each day is the most satisfactory method. It is for you to decide which pattern of study suits you best.

- If you decide to study for several hours at once, take short breaks of five or ten minutes every half hour or so. You will find that this method maintains a higher overall level of concentration.

Before you begin a detailed reading of the text, familiarise yourself with the general layout of the material. Have a look at the course contents list at the front of the book and flip through the pages to get a general impression of the way the subject is dealt with. You will find that there is space on the pages to make comments alongside the text as you study—your own notes for highlighting points that you feel are particularly important. Indicate in the margin the points you would like to discuss further with a tutor or fellow student. When you come to revise, these personal study notes will be very useful.

∏   When you find a paragraph in the text marked with a symbol such as is shown here, this is where you get involved. At this point you are directed to do things: draw graphs, answer questions, perform calculations, etc. Do make an attempt at these activities. If necessary cover the succeeding response with a piece of paper until you are ready to read on. This is an opportunity for you to learn by participating in the subject and although the text continues by discussing your response, there is no better way to learn than by working things out for yourself.

We have introduced self assessment questions (SAQ) at appropriate places in the text. These SAQs provide for you a way of finding out if you understand what you have just been studying. There is space on the page for your answer and for any comments you want to add after reading the author's response. You will find the author's response to each SAQ at the end of each part. Compare what you have written with the response provided and read the discussion and advice.

At intervals in the text you will find a Summary and List of Objectives. The Summary will emphasise the important points covered by the material you have just read and the Objectives will give you a checklist of tasks you should then be able to achieve.

You can revise the Unit, perhaps for a formal examination, by re-reading the Summary and the Objectives, and by working through some of the SAQs. This should quickly alert you to areas of the text that need further study.

Contents

Study Guide

Experimental design, although in reality a fairly well developed subject area, is only now beginning to find favour with scientists. This is especially true of industrial chemists. Some reasons why this might now be happening are the growing availability of microcomputing power and the gradual realization that these designs have a lot to offer in terms of potential savings in time and effort. As such then, the principal aim of this text is to make the subject available to a wider audience since it has till now remained the sole province of statisticians. At the same time I hope to provide the reader with sufficient statistical background on the assumptions and limitations of the methods used in the analysis of the experiments.

Part 1 of this unit provides a revision of some of the statistical methods used in experimental design, and may be excess to your requirements if you have anything more than a rudimentary understanding of statistics.

Part 2 introduces randomization, replication and blocking, followed by some designs which have increasing degrees of complexity as far as blocking is concerned (latin squares, graeco-latin squares and so on). These are what I consider to be some of the more fundamental areas of experimental design and you are recommended to read this part. You should certainly make sure you are clear about randomization and running experiments in blocks.

Part 3 introduces designs with more than one factor (factorial designs) and mainly covers the two-level designs, which are useful for discovering whether variables are important. The emphasis here lies in discovering whether main effects and interactions are important, although I have also covered several methods for their analysis. Amongst those I have introduced is Daniel's technique of plotting effects and residuals on normal probability paper. This technique is extremely useful when no replication is possible or desired. It enables the experimenter to take full advantage of all the information available in complete factorial designs. It is therefore advised that anyone who potentially has more than one factor influencing a response should read this section.

Part 4 briefly covers fractional factorial designs, which are very useful exploratory designs when you do not know whether certain factors have any effect upon an experimental response. These designs have the advantage over complete factorial designs when a large number of factors has to be included in an initial investigation, since they use a relatively small number of experimental runs.

Part 5 of this unit covers response surface methodology. If you are interested in mapping out a response surface and possibly finding an optimum combination of factors, Part 5 may provide you with a reasonably efficient means of accomplishing this. However, response surface techniques are only really useful if you already know what the important factors are and also if the response and factors vary in a continuous manner. I have included a brief introduction to matrix algebra which provides the easiest solutions in response surface techniques. Even if you have little idea about matrix algebra you should benefit quite a lot from working through the examples and intext questions, since it can provide a basis for working out the best designs for different experimental situations. However, I have structured the sections in such a way that you can miss out the discussion on their manipulation if you feel reasonably happy with them.

There are many software packages available to set up and analyse the results of these experiments. However, there is a tendency for the authors of these packages to produce a lot of output without too much explanation as to what it all means. Part 5 also introduces the basic models on which the designs are based, working through some of the calculations and then explaining what some of the terms which you might meet in typical output actually mean. You may then be able to decide which of the many available designs you should use to most efficiently map your response surfaces, either as contour diagrams or as the very attractive 3-D response surface plots. Part 5 also includes some fully worked examples which will be particularly important in providing you with some experience of the designs without necessarily having the wherewithal to construct and analyse them. Perhaps you will then be able to decide what you would like your experimental designs to achieve and therefore what you would like your software to do.

Computer Software

There are many commercial packages available which can calculate the statistics required for analysis of experiments. Many of these packages will apply the methods detailed in this text. However, few are directed specifically towards experimental design. Most tend to concentrate on producing statistics which are only really relevant to passively accumulated data. Also new software packages are continually appearing on the market. Therefore, it is not my intention to present a comprehensive list of packages, but instead to present some of the packages that I have either personally come into contact with or have read reviews of. You may of course know and love others. Inclusion of packages in this guide does not imply that I endorse or have any preference for them, and indeed I am sorry if I have offended anyone by not including their favourite package.

SPSSx/SPSSpc
This is a very powerful package with many routines for parametric and non-parametric analysis, available on both personal computers (PC) and mainframe. However it seems rather limited as far as experimental design is concerned, unable for example to produce 3-D response surface plots.

SAS
I have only seen a mainframe version of this. It seems very powerful, and handles all common statistical routines with good 3-D plots of response surfaces.

SYSTAT
This is available for the IBM PC and compatibles, copes with most types of statistical tests (parametric and non-parametric) and has 3-D plots.

UNISTAT III
Extremely comprehensive package available for the IBM PC and

compatibles. It has a good data processor based on a spread-sheet. The programme covers a large number of statistical methods of analysis including 3-D response surface plots.

ULTRAMAX
Optimization package for both the mainframe and IBM AT. This can optimize quadratic response surfaces in process development as a substitute for evolutionary operation (EVOP).

QSCA
This package is directed specifically towards experimental design and analysis without too much by way of other statistical methods. It is available for the IBM PC and compatibles and can design and analyse complete factorial, fractional factorial, and complete and incomplete blocked designs.

2^nFACTGRAPH
A spreadsheet programme to analyse complete and fractional factorial designs available for IBM PC and compatibles. It is essentially a template for Lotus 1-2-3 users which can estimate the importance of main and interactive effects, plot them on graphs and calculate residual errors for model checking.

XSTAT
This is available for IBM PC and compatibles. It is a true experimental design package which can design, randomize and analyse experiments with useful tips as you go along. Amongst others it develops one-factor designs, blocked designs, fractional and complete factorial designs, Box–Wilson, Box–Behnken and Plackett–Burmann designs. However, it does not produce 3-D response surface plots but instead adopts a contour map approach.

EXPERtIMENTAL DESIGN
This is available for the IBM PC and compatibles. It is an expert system which guides the user through the choice of designs for particular situations.

STATGRAPHICS
This package (for IBM PC and compatibles) is very good at the design and analysis of experiments, it produces a lot of statistics on the results (both non-parametric and parametric) and can be used to give 3-D response surface plots.

Bibliography

There are many books dealing with topics which are of interest in this Unit, and these approach their subject matter from a variety of points of view. Although I have tried to steer clear of original articles, I have included one or two which I feel could be of great benefit to you if you can obtain them.

1. C.K. Bayne and I.B. Rubin, *Practical Experimental Designs and Optimization Methods for Chemists*, VCH, 1986.

2. G.E.P. Box, W.G. Hunter and J.S. Hunter, *Statistics for Experimenters; An Introduction to Design, Data Analysis and Model Building*, Wiley, 1978.

3. C. Daniel, Use of half-normal plots in interpreting factorial two-level experiments, *Technometrics* 1959, 1, 311–341.

4. O.L. Davies, Ed. *The Design and Analysis of Industrial Experiments*, Longman, 1956.

5. N.R. Draper and H. Smith, *Applied Regression*, Wiley, 1981.

6. O.L. Davies and P.L. Goldsmith, *Statistical Methods in Research and Production*, Longman, 1984.

7. J.C. Miller and J.N. Miller, *Statistics for Analytical Chemists*, Ellis Horwood, 1984.

8. D. McCormick and A. Roach, *Measurement, Statistics and Computation*, ACOL, Wiley, 1987.

9. R.H. Myers, *Response Surface Methodology*, Allyn and Bacon, 1971.

10. R.L. Plackett and J.P. Burman, The design of optimum multifactorial experiments, *Biometrika* 1946, 33, 305–325.

1. Basic Statistics: A Review

The purpose of this introductory section is to ensure that you are able to apply a number of basic statistical tests which will be used later when examining various aspects of experimental design.

∏   Do you know what is meant by the terms accuracy, precision, standard deviation, random error and systematic error?

If your answer is "no", then you are advised to read the ACOL text on 'Measurement, Statistics and Computation'. If, however, your answer is "yes" then hopefully you should find this section quite straightforward as most of it is in the form of revision exercises. Please be aware, however, that towards the end of the chapter new material will be introduced.

1.1. ERRORS

Ideally you would like the result of any experiment to give a true response. In practice, however, there are three types of error which may affect an analytical result.

∏   Can you recall the names of these types of error?

You may have remembered the following:

(a) Gross errors, which require the experiment to be restarted, for example, the malfunctioning of an instrument, the realization that a reagent was badly contaminated, or that a vital step had been omitted in the analytical procedure.

(b) Systematic errors (or bias), which tend to give results that are always higher or lower than the true value. This could arise, for example, from someone who is unable to judge sharp colour changes in visual titrations, resulting in a slight over-reading of the end-point.

(*c*) *Random errors*, which arise from individual results falling on both sides of the true response. These errors are the slight variations that occur when successive measurements are made by the same person under as nearly identical conditions as possible.

You should realise that data obtained from a number of results is of limited use in itself. It is only by comparing the sets of results with a true value or with other sets of data that it is possible to show whether an analytical method is accurate and/or precise and is therefore in some way superior or inferior to another method.

∏   If a determination is accurate does it imply that it is free from systematic error?

Since accuracy may be defined as the closeness of a result to the *true* or *accepted* value then an accurate result should be free from a significant systematic error. Notice that the word "significant" has been included. Hence you should be able to appreciate that a significance test between results will tell you whether there is a systematic error (bias) or if the variation between results may be attributed to random error.

To illustrate this point Fig. 1.1 shows the results obtained for three methods (A, B and C). The results for method A are fairly similar but are all well above the true value. Thus method A is precise but is influenced by a positive systematic error. Method B, on the other hand, gives results which are rather more spread out. This method is therefore less precise than method A. However, the average result for method B comes out somewhere fairly close to the true value. Method B is therefore relatively free of systematic error. Clearly method C is both precise and accurate since the results are very similar and average out very close to the true value. One of the commonest methods of determining precision for a small set of data when it is expected to be distributed about the data in a *normal* or *gaussian* manner is the *standard deviation* (*s*), which is given by:

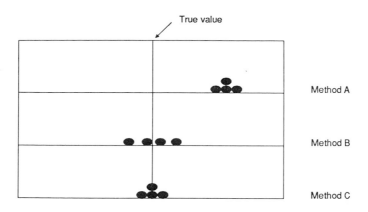

Fig. 1.1. *Effect of random and systematic errors upon analytical results*

$$s = \sqrt{\frac{\Sigma(\bar{y} - y_i^2)}{N - 1}}$$

in which y_i is one of the results from N samples and $\bar{y}$ is the sample average. However, having an estimate of standard deviation does not really allow you to make decisions as to whether one method is more precise or accurate than another.

∏   Can you recollect the names of the two usual methods you could use for comparing mean values and precisions in two data sets?

You may have remembered that Student's t-test can be used either for comparing the sample mean $\bar{y}$ with the true mean μ, or in a modified form can be used to determine whether there is a significant difference between the mean of one set of results, $\bar{y}_1$, and the mean of a second set of data, $\bar{y}_2$. If, however, you wish to compare the precisions of two different methods you would use the variance ratio test (F-test). Let us now examine the t-test in some detail, and hopefully refresh your memory by going through a few worked examples.

After answering the following SAQ, read the author's response in the section beginning on page 36.

SAQ 1.1a   The following results were obtained by two analysts using a new method for the determination of nickel in a standard reference alloy containing 6.25% nickel.

Analyst A	Analyst B
6.16	6.50
6.41	6.54
6.23	6.52
6.28	6.56

Mark the following statements with a cross ($\times$) if wrong, and a tick ($\checkmark$) if correct:

(*i*) The results obtained by analyst A are more precise than those of analyst B;

(*ii*) Since the results of analyst A fall on both sides of the true value, they could not be attributed to random error;

(*iii*) The range of results obtained by analyst A is 0.25;

(*iv*) It appears that analyst B's results are subject to a systematic error;

(*v*) The standard deviation of the results obtained by analyst A is $0.105_5\%$;

(*vi*) The variance of the results of analyst A is $\sqrt{0.105_5} = 0.324_8$

1.2.   THE *t*-TEST

The *t*-test is perhaps widely used by scientists because the test can be applied to a relatively small number of samples. One of the principal applications of the *t*-test is to compare an experimental mean $\bar{y}$ with a known true value (μ). You should remember to use the greek letters μ and σ for the population mean and population standard deviation respectively, whereas the corresponding terms for the sample are $\bar{y}$ and s. Consider first of all a specific example of the use of the *t*-test in which the true mean (μ) is known.

The specification for a certain alloy requires it to have 2.30% of vanadium. Ten replicate analyses of the alloy gave a mean vanadium content of 2.33% with a standard deviation of 0.02. Test at the $P = 0.01$ (1%) and $P = 0.05$ (5%) probability levels whether the vanadium content differs from the required specification.

The first step in the calculation is to evaluate t, which is given by:

$$t = \frac{(\bar{y} - \mu)\sqrt{N}}{s}$$

where $\bar{y} = 2.33\%$ and $\mu = 2.30\%$, $s = 0.02$ and N (the total number of replicate analyses) = 10.

$$t = \frac{(2.33 - 2.30)\sqrt{10}}{0.02}$$

$$= 4.74$$

∏   In this example do you use a one-tailed or a two-tailed test?

Well, I hope you answered "two-tailed" because you were asked whether the alloy differs from the required specification, which implies it may be above or below the specified value. Therefore, the test is two-tailed. If, however, the question had stated "test whether the vanadium content is higher than the specified value", then you would use a one-tailed test. You should also be aware of the number of degrees of freedom to use. In this example the number of degrees of freedom (*d.f.*) is given by ($N - 1$) where N is the number of replicate analyses. One *d.f.* has effectively been used up when calculating the average. Therefore, *d.f.* $= N - 1 = 10 - 1 = 9$.

Degrees of freedom (*d.f.*)	Probability level 0.05	0.025	0.01	0.005
1	6.31	12.71	31.82	63.66
2	2.92	4.30	6.97	9.92
3	2.35	3.18	4.54	5.84
4	2.13	2.78	3.75	4.60
5	2.02	2.57	3.37	4.03
6	1.94	2.45	3.14	3.71
7	1.90	2.36	3.00	3.50
8	1.86	2.31	2.90	3.36
9	1.83	2.26	2.82	3.25
10	1.81	2.23	2.76	3.17
11	1.80	2.20	2.72	3.11
12	1.78	2.18	2.68	3.05

Fig. 1.2.   *The value of t associated with a given area in the tail of a t distribution and given degrees of freedom (d.f.)*

The *t* distributions given above for 9 *d.f.* read as follows:

$$P = 0.05 \ (5\%) \quad 0.025 \ (2.5\%) \quad 0.01 \ (1\%) \quad 0.005 \ (0.5\%)$$
$$t = 1.83 \qquad\qquad 2.26 \qquad\qquad 2.82 \qquad\qquad 3.25$$

∏   What is the value of *t* when *d.f.* = 9 at the *P* = 0.05 (5%) probability level for a two-tailed test?

If you thought the *t* value was 2.26 then you obviously know how to use the *t*-tables. Remember, for a two-tailed test you double the value of *P* for a one-tailed test to obtain the probability level. For a one-tailed test, however, you merely need the *t* value for a given number of degrees of freedom at the stated probability.

∏   For a one-tailed test what would be the *P* = 0.01 (1%) probability value of *t* when *d.f.* = 9 ?

The answer in this case is 2.82. If you did not get this value, please read this section again carefully.

You now have to test the significance of this value to decide whether or not to reject the null hypothesis that the alloy is within specification. As the calculated value of *t* (4.74) is greater than the tabulated *t* value of 3.25, the *P* = 0.01 value for a two-tailed test, the probability of obtaining the difference between $\bar{y}$ and μ of 0.03% is less than 1 in 100. Thus, there is a significant difference between the analytical mean of vanadium, 2.33, and the specification value of 2.30 at the $P = 0.01$ probability level. You may have come across the term "null hypothesis" in your previous study of statistics, and may have been confused by it. Putting it simply, if the calculated (or observed) value of t is less than the tabulated value you can say that the null hypothesis ($H_0 : \bar{y} = \mu$) is retained and there is no evidence of a systematic error (bias) in the method. If, however, the calculated value of t is greater than the tabulated t value you can reject the null hypothesis and accept an alternative hypothesis ($H_1: \bar{y} \neq \mu$) instead. There is, therefore, a significant systematic error or bias in the procedure. In practice a probability level of 0.05 or 0.01 is usual. If you reject the null hypothesis on the basis of a probability level of 0.05, there are only five chances in a hundred of being wrong and you are therefore 95% confident that you have made the correct decision.

1.2.1. Comparison of the Means of Two Samples

So far we have only discussed the application of the t-test when the true mean (μ) is known. Frequently, however, you may wish to compare the results of a new analytical method with those of an established (reference) procedure. In this case you apply the t-test to compare two sample means, $\bar{y}_1$ and $\bar{y}_2$. The expression for t is given by:

$$t = \frac{\bar{y}_1 - \bar{y}_2}{s_p \sqrt{(1/N_1 + 1/N_2)}}$$

where N_1 and N_2 are the number of data points in each group and s_p is termed the pooled standard deviation, which is calculated from the two sample standard deviations, s_1 and s_2, by:

$$s_p = \sqrt{\frac{(N_1 - 1)s_1^2 + (N_2 - 1)s_2^2}{N_1 + N_2 - 2}}$$

Consider the following example:

The percentage yields of the reaction products of two methods, A and B, were determined. Five identical samples were used under the same conditions. The results are tabulated as follows:

	% yield	Standard deviation
Method A	25.8	0.51
Method B	25.1	0.31

Is method A better than method B? First you have to work out the pooled standard deviation for the following data:

$$N_A = N_B = 5, \; s_A = 0.51, \; s_B = 0.31$$

$$s_p = \sqrt{\frac{4 \times 0.51^2 + 4 \times 0.31^2}{8}}$$

$$s_p = 0.422$$

$$t = \frac{25.8 - 25.1}{0.422 \sqrt{(1/5 + 1/5)}} = \frac{0.7}{0.422 \times 0.632}$$

$$t = 2.62$$

∏   Do you use a one-tailed or a two-tailed test in this example?

It is reasonable to assume that *better* means a higher % yield so you use a one-tailed test.

∏   How many degrees of freedom are there?

The number of degrees of freedom is given by $N_A + N_B - 2$ and is therefore equal to eight. Looking up the t-tables for eight degrees of freedom you should find:

$$P = 0.05 \quad (5\%) \quad 0.01 \quad (1.0\%)$$
$$t = 1.86 \qquad\qquad 2.90$$

You use the null hypothesis that the mean percentage yields are the same for both method A and method B. Now the observed value of 2.62 is greater than the $P = 0.05$ value (1.86) but smaller than the $P = 0.01$ value (2.90). So you can say that at the 5% probability level method A is better than method B, but are unable to make this claim at the 1% probability level. Expressed in a different way, you can be 95% certain that method A is better than method B but cannot be 99% certain. However, the value 2.62 is closer to 2.90 than 1.86, so the statement that you are almost 99% certain that method A is superior to method B is justified.

1.2.2. The Paired *t*-Test

A situation can arise when you wish to compare two analytical methods where the samples tested are of varying composition. Let us examine the following example:

Two different methods, A and B, were tested for the analysis of five different manganese compounds. These are shown in the following table:

Sample	1	2	3	4	5
Method A	16.8	7.6	18.2	32.5	19.7
Method B	17.1	7.9	17.8	32.3	20.2

In testing whether there is a significant difference between methods A and B, the appropriate null hypothesis would seem to be that the two methods are the same. I hope you realise it would be a nonsense to work out the means and standard deviations and calculate t as in the previous example since any differences between the methods will be swamped by differences in the samples. The key word here is "different" manganese compounds.

In this case you apply the *paired t-test*. Let us work through the above example. We have to work out the differences (d) between each pair of results and then calculate $\bar{d}$, the mean of the differences. We then evaluate the standard deviation of the differences, s_d. These are shown in the table below:

Method A	Method B	d	$(\bar{d} - d_i)$	$(\bar{d} - d_i)^2$
16.8	17.1	+0.3	−0.2	0.04
7.6	7.9	+0.3	−0.2	0.04
18.2	17.8	−0.4	0.5	0.25
32.5	32.3	−0.2	0.3	0.09
19.7	20.2	+0.5	−0.4	0.16
	Σd	=0.5	$\Sigma (\bar{d} - d_i)^2$	= 0.58
	$\bar{d}$	= 0.1		

The value of t is calculated from the equation

$$t = \frac{\bar{d}\sqrt{N}}{s_d}$$

and the number of degrees of freedom is $N - 1$.

∏   Calculate the value of t for the above example.

The first thing you should have done is evaluate s_d (the standard deviation of differences) in the same way as you would have for an ordinary standard deviation (summing the squared deviations and dividing by the *d.f.*). I worked it out to be:

$$s_d = \sqrt{\frac{0.58}{4}} = 0.38$$

Then using the above equation you should have calculated t;

$$t = \frac{0.10\ \sqrt{5}}{0.38} = 0.588$$

For a two-tailed test with four *d.f.* (5 − 1) the tabulated value of *t*, at $P = 0.05$, is 2.78. The calculated value of *t* is much less than this 5% *t* value. Hence there is no significant difference between the methods at this probability level. That is to say, the null hypothesis can be retained and you may say that the two methods do not give significantly different values for the manganese concentrations of different samples.

SAQ 1.2a   In a paired *t*-test comparing two methods, X and Y, in which twelve samples were analysed, a *t* value of 2.20 was calculated. From this information and from *t*-tables select with a circle the correct answer to each of the following questions:

(*i*) How many degrees of freedom are there? (11 or 12 or 13).

(*ii*) For a two-tailed test what is the value of *t* at the $P = 0.05$ level given $N = 12$? (1.78 or 1.80 or 2.18 or 2.20).

(*iii*) Does the calculated value of *t* (2.20) lead us to the conclusion that method Y gives significantly higher results than method X at the $P = 0.05$ probability level? (No, or they are the same, or Yes).

SAQ 1.2b | A chemical manufacturer wishes to test a new route (B) for the synthesis of an organic compound. The original method of synthesis (A) has been used by the company for many years. Purely on the statistical evidence of the yields given below, determine whether it is sensible for the manufacturer to change to method B for the production of the organic compound.

	Number of results	Yield (%) of the organic	Standard deviation
Method A	5	72.3	0.36
Method B	6	72.9	0.33

1.3. THE F-TEST

We stated previously that if we wished to compare the precisions of two different methods we would use the variance ratio or F-test. However, the F-test will also be used extensively later in this book, when variances deliberately introduced by changing treatments will be compared with variances due to random error. The value of F is calculated from the equation:

$$F = s_A^2/s_B^2 = V_A/V_B$$

where s^2 = (the standard deviation)^2 = the variance (V).

For the two variances to be significantly different the ratio has to be greater than unity, and it is usual to put the larger value of V as the numerator and the smaller V value as the denominator. However, in some circumstances you may wish to test, for example, whether the precision of one method is significantly greater than that of a second method, in which case you make a ratio of the variance for the first method to the variance of the second method. Obviously if it subsequently turns out that the ratio has a value less than unity there is no point in looking up the F-tables.

∏   When does the F ratio = 1?

Clearly F can only equal unity when $s_A^2 = s_B^2$, that is to say when the precisions of methods A and B are the same.

An application of the F-test is shown in the following example.

The standard deviation from a set (A) of 13 determinations is 0.290 and the standard deviation from another set (B) of 10 determinations is 0.642.

∏   Is the precision of the first method significantly greater than that of the second method?

The value of F is given by

$$F = \frac{(0.642)^2}{(0.290)^2} = 4.90$$

∏   Do you recollect how to use the F-tables?

In F-tables there is a row at the top of the table in addition to the column at the left-hand side of the table, and both give numbers of degrees of freedom ($d.f._A$ and $d.f._B$ respectively). The number of degrees of freedom associated with each variance is simply $N - 1$, where N is the number of samples or determinations. In this example, in which $d.f._B = 10 - 1 = 9$, you follow the values on the top row until you have $d.f. = 9$, because this is the number of determinations for the less precise method which forms the numerator in the F ratio. The second value for the degrees of freedom is $N_A - 1$, where N_A is the number of determinations for method A and is therefore 12. This is the more precise and forms the denominator in the F ratio. Arranging the variances in this way ensures an F ratio of greater than unity. In this example $N_A = 13 - 1 = 12$, and you follow the column on the left-hand side of the F-tables down until you reach $d.f. = 12$.

Using the appropriate $d.f.$ values, F values can be read from the F-tables. Fig. 1.3a should help you to read these if you do not know how to do this already.

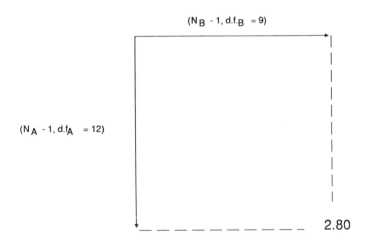

Fig. 1.3a. *In F-tables the degrees of freedom for the numerator variance are read horizontally and the degrees of freedom for the denominator variance are read vertically (values for P = 0.05)*

For the above example the tabulated F value at a probability of $P = 0.05$ (5%) is 2.80. Here you are dealing with a one-tailed test since you are asked whether the precision of method A is significantly greater than that of method B. The calculated value of F (4.90) exceeds the tabulated F value at the $P = 0.05$ probability level for 9 and 12 $d.f.$ (2.80), and is also greater than the $P = 0.01$ value of 4.39. Therefore, you can say that the

$df._1$	1	2	3	4	5	6	7	8	9	10
$df._2$										
1	161. 45	199. 50	215. 71	224. 58	230. 16	233. 99	236. 77	238. 88	240. 54	241. 88
2	18. 51	19. 00	19. 16	19. 25	19. 30	19. 33	19. 35	19. 37	19. 38	19. 40
3	10. 13	9. 55	9. 28	9. 12	9. 01	8. 94	8. 89	8. 85	8. 81	8. 79
4	7. 71	6. 94	6. 59	6. 39	6. 26	6. 16	6. 09	6. 04	6. 00	5. 96
5	6. 61	5. 79	5. 41	5. 19	5. 05	4. 95	4. 88	4. 82	4. 77	4. 74
6	5. 99	5. 14	4. 76	4. 53	4. 39	4. 28	4. 21	4. 15	4. 10	4. 06
7	5. 59	4. 74	4. 35	4. 12	3. 97	3. 87	3. 79	3. 73	3. 68	3. 64
8	5. 32	4. 46	4. 07	3. 84	3. 69	3. 58	3. 50	3. 44	3. 39	3. 35
9	5. 12	4. 25	3. 86	3. 63	3. 48	3. 37	3. 29	3. 23	3. 18	3. 14
10	4. 96	4. 10	3. 71	3. 48	3. 33	3. 22	3. 14	3. 07	3. 02	2. 98
11	4. 84	3. 98	3. 59	3. 36	3. 20	3. 09	3. 01	2. 95	2. 90	2. 85
12	4. 75	3. 89	3. 49	3. 26	3. 11	3. 00	2. 91	2. 85	2. 80	2. 75
13	4. 67	3. 81	3. 41	3. 18	3. 03	2. 92	2. 83	2. 77	2. 71	2. 67
14	4. 60	3. 74	3. 34	3. 11	2. 96	2. 85	2. 76	2. 70	2. 65	2. 60
15	4. 54	3. 68	3. 29	3. 06	2. 90	2. 79	2. 71	2. 64	2. 59	2. 54

Fig. 1.3b. *The values of F at a probability level of 0.05*

$df._1$	1	2	3	4	5	6	7	8	9	10
$df._2$										
1	647. 79	799. 50	864. 16	899. 58	921. 85	937. 11	948. 22	956. 66	963. 28	968. 63
2	38. 51	39. 00	39. 17	39. 25	39. 30	39. 33	39. 36	39. 37	39. 39	39. 40
3	17. 44	16. 04	15. 44	15. 10	14. 88	14. 73	14. 62	14. 54	14. 47	14. 42
4	12. 22	10. 65	9. 98	9. 60	9. 36	9. 20	9. 07	8. 98	8. 90	8. 84
5	10. 01	8. 43	7. 76	7. 39	7. 15	6. 98	6. 85	6. 76	6. 68	6. 62
6	8. 81	7. 26	6. 60	6. 23	5. 99	5. 82	5. 70	5. 60	5. 52	5. 46
7	8. 07	6. 54	5. 89	5. 52	5. 29	5. 12	4. 99	4. 90	4. 82	4. 76
8	7. 57	6. 06	5. 42	5. 05	4. 82	4. 65	4. 53	4. 43	4. 36	4. 30
9	7. 21	5. 71	5. 08	4. 72	4. 48	4. 32	4. 20	4. 10	4. 03	3. 96
10	6. 94	5. 46	4. 83	4. 47	4. 24	4. 07	3. 95	3. 85	3. 78	3. 72
11	6. 72	5. 26	4. 63	4. 28	4. 04	3. 88	3. 76	3. 66	3. 59	3. 53
12	6. 55	5. 10	4. 47	4. 12	3. 89	3. 73	3. 61	3. 51	3. 44	3. 37
13	6. 41	4. 97	4. 35	4. 00	3. 77	3. 60	3. 48	3. 39	3. 31	3. 25
14	6. 30	4. 86	4. 24	3. 89	3. 66	3. 50	3. 38	3. 29	3. 21	3. 15
15	6. 20	4. 77	4. 15	3. 80	3. 58	3. 41	3. 29	3. 20	3. 12	3. 06

Fig. 1.3c. *The values of F at a probability level of 0.025*

variance of method B is significantly higher than the variance of method A at the 1% level and that method A is more precise.

Quite often you wish to test whether the variances of two methods differ significantly and not whether one is specifically higher than another. In such a case you should use a two-tailed test. Therefore, one-tailed probability

$d.f._1$	1	2	3	4	5	6	7	8	9	10
$d.f._2$										
1	4052. 2	4999. 5	5403. 4	5624. 6	5763. 6	5859. 0	5928. 4	5981. 1	6022. 5	6055. 8
2	98. 50	99. 00	99. 17	99. 25	99. 30	99. 33	99. 36	99. 37	99. 39	99. 40
3	34. 12	30. 82	29. 46	28. 71	28. 24	27. 91	27. 67	27. 49	27. 35	27. 23
4	21. 20	18. 00	16. 69	15. 98	15. 52	15. 21	14. 98	14. 80	14. 66	14. 55
5	16. 26	13. 27	12. 06	11. 39	10. 97	10. 67	10. 46	10. 29	10. 16	10. 05
6	13. 75	10. 92	9. 78	9. 15	8. 75	8. 47	8. 26	8. 10	7. 98	7. 87
7	12. 25	9. 55	8. 45	7. 85	7. 46	7. 19	6. 99	6. 84	6. 72	6. 62
8	11. 26	8. 65	7. 59	7. 01	6. 63	6. 37	6. 18	6. 03	5. 91	5. 81
9	10. 56	8. 02	6. 99	6. 42	6. 06	5. 80	5. 61	5. 47	5. 35	5. 26
10	10. 04	7. 56	6. 55	5. 99	5. 64	5. 39	5. 20	5. 06	4. 94	4. 85
11	9. 65	7. 21	6. 22	5. 67	5. 32	5. 07	4. 89	4. 74	4. 63	4. 54
12	9. 33	6. 93	5. 95	5. 41	5. 06	4. 82	4. 64	4. 50	4. 39	4. 30
13	9. 07	6. 70	5. 74	5. 21	4. 86	4. 62	4. 44	4. 30	4. 19	4. 10
14	8. 86	6. 51	5. 56	5. 04	4. 69	4. 46	4. 28	4. 14	4. 03	3. 94
15	8. 68	6. 36	5. 42	4. 89	4. 56	4. 32	4. 14	4. 00	3. 89	3. 80

Fig. 1.3d. *The values of F at a probability level of 0.01*

values at $P = 0.025$ have been included in the F-tables. These will, on doubling, give you the $P = 0.05$ (5%) values of F for a two-tailed test.

1.3.1. Alternative t- And F-Tests

As a result of attempting the solution of many statistical problems you may well be tempted to ask: "virtually all the questions that I have been set rely on at least five replicate determinations. What about the more realistic situation, where I have only three, or at very most, four replicated results? Also, are there any tests available for a small number of experiments in which the data do not necessarily behave in a gaussian manner?" You will no doubt be pleased to hear that there are, and that in general they are very easy to work out.

One alternative F-test is the F_R test, which compares the ranges of two sets of results:

$$F_R = W_1/W_2$$

where W_1, the range of one set of results, is usually greater than the range W_2 of the other set. Consider the following example:

The end-points obtained by two analysts using the same titration method were as follows:

Analyst I   25.03,   25.08,   25.10,   25.14   ml
Analyst II   25.14,   25.18,   25.16,   25.12   ml

Using the ranges of the results test whether (*i*) there is a significant difference in precision between the analysts at the $P = 0.05$ probability level and (*ii*) the two analysts obtain significantly different mean results ($P = 0.05$).

The range W_1 for analyst I = 25.14 − 25.03 = 0.11 ml and for analyst II = 25.18 − 25.12 = 0.06 ml. To test whether there is a difference in precisions you apply the following equation:

$$F_R = W_1/W_2 = 0.11/0.06 = 1.83$$

Number of measurements in the numerator and denominator	One-tailed test	Two-tailed test
2	12. 7	25. 5
3	4. 4	6. 3
4	3. 1	4. 0
5	2. 6	3. 2
6	2. 3	2. 8
7	2. 1	2. 5
8	2. 0	2. 3
9	1. 9	2. 2
10	1. 9	2. 1

Fig. 1.3e. *Critical F_R values for one-tailed and two-tailed tests at P = 0.05*

From the F_R table given in Fig. 1.3e you can see that the critical value of $F_{R(4)}$ for a two-tailed test at $P = 0.05$ is 4.0. Since the calculated value of $F_R = 1.83$ is less than this tabulated value you may conclude there is no significant difference in the precision of the results obtained by two analysts.

Suppose you now wish to test whether the mean values for the two analysts are significantly different. To do this with a small number of results you can use an alternative t-test based on the T_d statistic which is given by:

$$T_d = \frac{2 \mid \bar{y}_1 - \bar{y}_2 \mid}{W_1 - W_2}$$

where $\bar{y}_1$ and $\bar{y}_2$ are the means of two sets of results. In our example:

$$T_d = \frac{2 \mid 25.09 - 25.15 \mid}{0.11 + 0.06} = \frac{0.12}{0.17} = 0.71$$

$N_1 = N_2$	T_d
2	3. 43
3	1. 27
4	0. 81
5	0. 61
6	0. 50
7	0. 43
8	0. 37
9	0. 33
10	0. 30

Fig. 1.3f. *Critical values for T_d at $P = 0.05$*

The tabulated value of T_d for four pairs of results = 0.81 ($P = 0.05$). The calculated value is less than this. Therefore, the null hypothesis is not rejected and you can say that the mean values obtained by the two analysts do not differ at this probability level.

The F_R and T_d tests differ from the usual F- and t-tests because they can only deal with examples when the number of results is the same in each case, i.e., $N_1 = N_2$. These tests based on ranges are useful in dealing with very small samples of data. Although they are perhaps not as rigorous as the t- and F-tests discussed earlier, the major advantages of these range tests are that the calculations are simple, they can be applied to very small data sets and do not assume normality of distribution.

SAQ 1.3a	A new method for the determination of dissolved oxygen in water was compared with the standard (Winkler) method. The following results were obtained for a river water sample.

New method: 62; 66; 63; 65% dissolved O_2
Standard method: 67; 66; 66; 67% dissolved O_2

Use the F-test and the F_R test to examine whether the precisions of the methods differ significantly?

At the present stage we have revised the application of t-tests either to compare a sample mean with a known or true value or to compare the means of two sets of data. We have also used the F-test to examine the relative precisions of two methods. It is, however, not an unrealistic situation to want to compare the means of more than two sets of results. You may, for example, wish to compare the performance of a number of analysts using the same method, e.g. three analysts using the same solutions for three replicate titrations. In such a case there are two likely sources of error, (i) the random error you always get with replicate measurements and (ii) the variation that may occur between the individual analysts.

To solve problems of this type a statistical method known as *Analysis of variance* may be used (often abbreviated to *ANOVA*).

1.4. ANALYSIS OF VARIANCE (*ANOVA*)

When you wish to compare more than two treatment or sample means, the null hypothesis to be tested is usually that the t treatment or sample means are the same, and the alternative hypothesis is that they are not. Analysis of variance (*ANOVA*) is a useful technique for making decisions about these hypotheses. In analysis of variance it is actually the variation in the t treatment or sample responses which is used to decide whether or not treatment effects are significant. The techniques are usually justified on the supposition that the data can be treated as random samples from t normal populations having the same variance, σ^2, and differing, if at all, only in their means.

The null hypothesis in this case is that the treatment means are not different, and that they are taken from a population of treatment means. On this basis the variance in the data may be assessed in two ways, one from between the treatment means and another from within the treatments. The analysis of variance will then determine whether these variances are the same size by means of an F-test. In *ANOVA* the F-test is one-tailed because you are only interested in whether the variance introduced into the data as a result of having different means is significantly greater than the random variation in the data. Therefore, the further apart the treatment means are, the more likely it is that they are significantly different. An *ANOVA* calculation is best illustrated by the use of a specific example. A general solution will then be developed subsequently.

1.4.1. One-factor *ANOVA* Example

The following values were obtained for trace amounts of Cd^{2+} from shoots of plants grown in soil which had been amended with different amounts of sewage sludge (50, 100 and 150 grams of sludge per kilogram of soil). Because of the complex nature of ion uptake by plant roots from soil and translocation into the shoots it does not automatically follow that increasing the amounts of metal-bearing sludge will increase the shoot Cd^{2+} concentrations. Four plants were taken from each soil. You wish to find out whether there is a significant difference between the mean plant cadmium values from the three soils at the $P = 0.05$ probability level. The results are given in the following table:

| | Grams of sludge added per kg soil | | |
	50 (A)	100 (B)	150 (C)
Shoot Cd conc. (μg g^{-1})	33.0	35.5	42.2
	33.4	33.5	38.8
	34.0	36.0	40.5
	33.6	35.0	36.5

Be warned, the calculation is quite lengthy but it should not be too difficult to follow. Let's go through the solution step by step.

In order to simplify the arithmetic it is a good idea to subtract a common value from each value of the data tabulated above. In this case I have decided to subtract the average value, 36.0, from each value. It does not matter which number is subtracted, it will not affect the final *ANOVA* from which we will make our inferences about the treatment means.

It is a clever move to subtract a number that will give zero values, you will soon see why this is true. So the first step is to subtract 36.0 and determine the sum of each column:

	Treatment A (50)	Treatment B (100)	Treatment C (150)
	− 3.0	−0.5	6.2
	− 2.6	−2.5	2.8
	− 2.0	0.0	4.5
	− 2.4	-1.0	0.5
Sum	−10.0	−4.0	14.0

Notice that each column has been summed. We now have to go through a series of stages, to calculate:

(*a*) the *Grand Total T*:

$$T = -10.0 - 4.0 + 14.0 = 0.0$$

obtained by adding the sum total of each individual column. This is the same as multiplying the average by the total number of observations.

$$T = N\bar{y}$$

where N is the total number of results (in this example, $N = 12$)

(*b*) the *Correction Factor* or *Correction for the Mean (CF)*:

$$CF = T^2/N$$

which in this case is

$$CF = 0/12 = 0.0$$

(*c*) the *Total (Corrected) Sum of Squares (TotalSS)*:

This is evaluated by squaring each result in the above table and then summing the column totals obtained. At this point it is usual to subtract the correction factor (*CF*) but since this is zero there would seem little point.

	Treatment A	Treatment B	Treatment C
	9.00	0.25	38.44
	6.76	6.25	7.84
	4.00	0.00	20.25
	5.76	1.00	0.25
Sum	25.52	7.50	66.78

Now, summing the column totals, the total corrected sum of squares is obtained:

$$TotalSS = 25.52 + 7.50 + 66.78 = 99.8$$

Although it is not intended to prove it here, it is possible to show that the total corrected sum of squares can be broken down into its components. These are (*i*) the variability introduced by the samples or treatments, called the *Between-treatment Sum of Squares (TreatmentSS)*, and (*ii*) the random variability, called the *Within-treatment* or *Residual Error Sum of Squares (ResidSS)*.

$$TotalSS = TreatmentSS + ResidSS$$

(*d*) the Between-treatment Sum of Squares (*TreatmentSS*):

This is calculated by squaring the column totals in the above table and then adding these together. This value is divided by the number of results in each column (in this example, 4). Finally the correction factor (*CF*) is subtracted, but as this is zero in this case it makes no difference.

$$TreatmentSS \ = \ \frac{(-10.0)^2 + (-4.0)^2 + (14.0)^2}{4} - CF$$

$$= \ \frac{100 + 16 + 196}{4} - 0.0$$

$$= \ 312/4 = 78$$

(*e*) the Residual Error Sum of Squares (*ResidSS*)

As shown above the *TreatmentSS* and *ResidSS* add up to the *TotalSS*. Therefore, simply subtracting the between-treatment sum of squares from the total corrected sum of squares will give the *ResidSS*.

$$ResidSS \ = \ 99.8 \ - \ 78 = 21.8$$

(*f*) The Degrees of Freedom (*d.f.*)

In any experiment the total number of degrees of freedom which are available is given by the total number of experiments. However, as discussed earlier, once an average has been calculated, one degree of freedom is lost. Therefore, the total number of degrees of freedom is given as $N - 1$, which in this example is $12 - 1 = 11$.

From here on it will be assumed that the degree of freedom associated with the average has already been accounted for.

$$\text{The between-treatment } d.f. \ = \ t \ - \ 1 \ = \ 2$$
$$\text{The residual error } d.f. \ = \ N \ - \ t \ = \ 9$$

where *t* is the number of treatments given as the number of columns (in this case = 3).

The number of degrees of freedom for the residual error sum of squares (*Resid d.f.*) is therefore the total number of experimental runs (N) minus the number of treatments (t). An average estimate of the residual error or within-treatment variance, known as the residual error or within-treatment mean square (*ResidMS*), is obtained by dividing the residual sum of squares by $N - t$.

$$ResidMS \; = \; ResidSS\,/Resid \; d.f. \; = \; 21.8/9 = 2.422$$

This is done on the assumption that each of the individual observations has been randomly taken from a population for the particular treatment. The residual error or within-treatment mean square therefore supplies an estimate of the residual error or within-treatment variance, σ^2, based on $N - t$ degrees of freedom.

The overall average $\bar{y}$ is defined as the sum of all the observations divided by the total number of observations (N). If there are no real differences between the treatment means then a second estimate of σ^2 can be obtained from the variation of the treatment means about the overall average. This estimate, known as the between-treatment mean square (*TreatmentMS*), is formed by dividing the between-treatment sum of squares (*TreatmentSS*) by the between-treatment degrees of freedom (number of treatments minus one).

$$TreatmentMS \; = \; TreatmentSS\,/Treatment \; d.f. \; = \; 78/2 = 39$$

On the null hypothesis that there are no real differences between the treatment means, there are now two estimates of σ^2 available. If the between-treatment mean square is much greater than the residual error or within-treatment mean square, there is little chance that the null hypothesis is correct and real differences between the treatments may have caused the between-treatment variation to be large.

(*g*) We now have sufficient data to set up what is called an Analysis of Variance Table, *ANOVA* Table.

The *ANOVA* Table is drawn up as shown below.

Source of variation	d.f.	SS	MS
Between-treatments	2	78	78/2 = 39
Residual error	9	21.8	21.8 = 2.422
Total	11	99.8	

(h) A one-tailed F-test is used to compare the two mean squares

$$F = 39/2.422 = 16.10$$

From the F-tables, the critical value of F at $P = 0.05$ for 2 and 9 d.f. is 4.25, above which the null hypothesis will be rejected. Since the observed F value is greater than this, it may be concluded that there is a significant difference between the shoot cadmium levels in plants grown in the three sludge-amended soils.

This has been a lengthy problem and several important features of the calculation will now be highlighted.

The first point to note in any *ANOVA* problem is to use as many significant figures as possible, do not "round off" values until the very last stage of the calculation. The reason for this should become apparent when you attempt the next SAQ.

Although the next points may seem obvious, they are still worth stressing. The total corrected sum of squares must be positive. It could be a very small value say, 0.000126, but is still positive. The between-treatment sum of squares must also be positive and numerically less than the total corrected sum of squares. If you think about it, squaring any real number results in a positive value, so it is not possible for any "sum of squares" terms to be negative. Finally you should be aware that we were investigating one factor, that is the cadmium levels of three different treatments. An *ANOVA* calculation involving just one factor (variable) is usually termed a *One-way* or *One-factor ANOVA*.

1.4.2. Analysis of the Residuals

Data in which the mean values for the treatments are different, but which
have the same variances, is said to be *homoscedastic*, as opposed to having
different variances, which is said to be *heteroscedastic*. These two situations
are shown in Fig. 1.4a and Fig. 1.4b respectively.

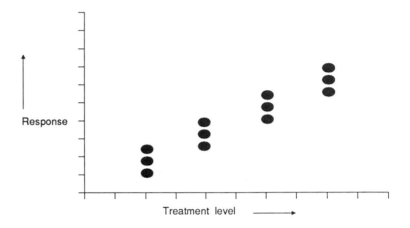

Fig. 1.4a. *Homoscedastic variation in which variance is constant with
increasing mean response*

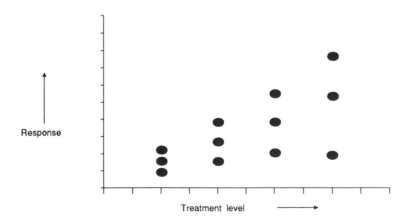

Fig. 1.4b. *Heteroscedastic variation in which variance increases with
mean response*

Unfortunately, analysis of variance is very susceptible to heteroscedasticity because it attempts to use a comparison of the estimates of variance from different sources to infer whether the treatments have a significant effect. If the data tends to be heteroscedastic it might be necessary to *transform* the results in some way to stabilize the variance and repeat the *ANOVA*. This is an example of what can be accomplished by retrospectively examining the residual variation about the treatment means.

Let us now examine the residuals obtained in the above example. The simplest way of obtaining these is to subtract the treatment average from the individual values. I have done this in the following table:

Treatment A response	average	residual
−3.0	−2.5	−0.5
−2.6	−2.5	−0.1
−2.0	−2.5	0.5
−2.4	−2.5	0.1

Treatment B response	average	residual
−0.5	−1.0	0.5
−2.5	−1.0	−1.5
0.0	−1.0	1.0
−1.0	−1.0	0.0

Treatment C response	average	residual
6.2	3.5	2.7
2.8	3.5	−0.7
4.5	3.5	1.0
0.5	3.5	−3.0

$\prod$ What distribution should these residuals follow?

You should expect the residuals to lie normally about the treatment average response. If this is not the case this may be evidence of some trend within the responses.

It was stated earlier that the treatment averages may differ but that they should all have the same variance, σ^2. Here, there is a clear trend of increasing variance about the treatment averages. This is a serious problem in *ANOVA*. One way of dealing with it is to transform the response so that the variances no longer change so drastically according to the average treatment response. Several transforms are available, and the one you use depends upon the exact relationship between the treatment averages and the variance. One commonly used technique is to take the square root of the response (before subtracting the arbitrary number or average response). This is often applied to situations in which the residuals follow a Poisson distribution. It is regarded as being a fairly mild transformation. Other methods include using the logarithm of the response or taking the reciprocal or reciprocal square root; the latter two being fairly strong transformations. In the subsequent *ANOVA* the residual variance is given one less degree of freedom because one has been used up in the transformation.

You can often test the effectiveness of such actions by making ratios of the largest to smallest variances before and after transformation.

$\prod$ What is the effect upon the variances of using the $\log_{10}$ of the shoot Cd^{2+} concentration? The original and transformed responses for the three treatments are given below:

Treatment A		Treatment B		Treatment C	
response	$\log_{10}$	response	$\log_{10}$	response	$\log_{10}$
33.0	1.519	35.5	1.550	42.2	1.625
33.4	1.524	33.5	1.525	38.8	1.589
34.0	1.531	36.0	1.556	40.5	1.607
33.6	1.526	35.0	1.544	36.5	1.562

Using the original data I found that the variance of treatment A (the smallest) was 0.173, and the variance of treatment C (the largest) was

5.927, which worked out to be a ratio of 34.26. On the other hand, when using the log $_{10}$ responses I found that the ratio fell to 24.95. Whether this has a significant effect upon the analysis of variance is for you to determine, but in any case it is likely to be more valid than before, even though the variances are still quite different.

1.4.3. *ANOVA* – The General Form

If results are arranged in columns and rows, as in *ANOVA*, there is a general way of representing this data. The results are shown in Fig. 1.4c for an experiment having t treatments or samples, each of which has r replications.

Treatments or samples

		1	2	j	t
Replications	1	y_{11}	y_{2}	y_{j1}	y_{t1}
	2	y_{12}	y_{22}	y_{j2}	y_{t2}
	.	.	.	.	.
	.	.	.	.	.
	k	y_{1k}	y_{2k}	y_{jk}	y_{tk}
	.	.	.	.	.
	.	.	.	.	.
	r	y_{1r}	y_{2r}	y_{jr}	y_{tr}

Fig. 1.4c. *Arrangement of results for ANOVA*

Hence we can say that y_{25} refers to the fifth measurement on the second sample.

∏ What does y_{jk} denote?

This is simply the kth measurement on the jth sample.

Let us now attempt to produce the general formulae for the one-factor *ANOVA* calculation. Initially the following are defined:

(*i*) The total number of measurements (N)

$$N = t \times r$$

(*ii*) The sum of measurements on the jth sample or treatment

$$= \sum_{k=1}^{k=r} y_{jk}$$

(*iii*) The grand total $= T = \sum_{j=1}^{j=t} \sum_{k=1}^{k=r} y_{jk}$

∏   How can you represent the correction factor (*CF*) using this notation?

You may write the *CF* as T^2/N, but it can also be written as $N\bar{y}^2$. This is no different from the symbols used when solving the previous worked example. I hope you agree that this was quite simple, but now you are going to tackle a harder problem and this requires your total attention.

∏   How do you represent the total corrected sum of squares as a general expression?

Remember at first you have to square every measurement, which may be written as y_{jk}^2. Next you have to sum each of the squared values in each column. This becomes

$$\sum_{k=1}^{k=r} y^2_{jk}$$

If you find this difficult to follow consider this example. Assume you have just one sample or treatment, so then $t = 1$. If there are three replicate values, then $r = 3$. You then have the following values:

2 (y_{11})   squaring these we get   4 (y_{11})^2
3 (y_{12})   9 (y_{12})^2
1 (y_{13})   1 (y_{13})^2
                                                    _____

$\sum_{k=1}^{k=r} y^2_{jk}$ =   14

Don't forget the symbol

$$\sum_{k=1}^{k=3} y^2 jk$$

means the sum of the y_{jk}^2 terms from $k = 1$ to $k = 3$. Next, sum together the totals of each column, so you have to sum each term from $j = 1$ to $j = t$. This gives a term which is called the *Crude Sum of Squares (CrudeSS)*:

$$CrudeSS = \sum_{j=1}^{j=t} \sum_{k=1}^{k=r} y^2 jk$$

However, so far the correction for the mean has not been taken into account. Subtraction of this term from the *CrudeSS* leaves the total corrected sum of squares (*TotalSS*):

$$TotalSS = \sum_{j=1}^{j=t} \sum_{k=1}^{k=r} y^2\, jk - T^2/N$$

The between-treatment sum of squares is obtained from (the sum of measurements)^2 which is given by

$$TreatmentSS = \sum_{j=1}^{j=t} \left(\sum_{k=1}^{k=r} y_{jk} \right)^2$$

which is divided by the number of replications (r) and the correction for the mean is subtracted.

∏   Can you write down the formula for the between-treatment sum of squares (*TreatmentSS*)?

It is hoped that you wrote

$$TreatmentSS = \frac{\sum_{j=1}^{j=t} \left(\sum_{k=1}^{k=r} y_{jk} \right)^2}{r} - T^2/N$$

Although it is often of benefit to the simplicity of the analysis, it is not always necessary to have the same number of replicates (r) for each treatment. This can be indicated in any of the above formulae by the subscript k next to the t, to give t_k.

Having worked out the values for both the *TotalSS* and *TreatmentSS* you can obtain the residual error sum of squares simply by subtracting *TreatmentSS* from *TotalSS*.

You may find the Σ notation a little difficult at first. If so, there's a useful tip that does not appear in most statistics books. Always treat these Σ values from right to left. So if we have

$$\sum_{j=1}^{j=t} \left(\sum_{k=1}^{k=r} y_{jk} \right)^2$$

Always evaluate

$$\left(\sum_{k=1}^{k=r} y_{jk} \right)^2$$

first and then proceed to calculate

$$\sum_{j=1}^{j=t} \left(\sum_{k=1}^{k=r} y_{jk} \right)^2$$

SAQ 1.4a	Four students were each asked to perform three replicate titrations using the same titrimetric procedure. The end-points obtained by each student are shown in the following table. $\longrightarrow$

Student	A	B	C	D
	25.03	24.95	25.13	25.16
Titres	25.09	24.90	25.20	25.25
(ml)	25.06	24.89	25.08	25.10

Determine whether there is a significant difference (at the $P = 0.05$ level) in the mean titration value obtained by each student. Present your final results in an *ANOVA* table.

1.4.4. Tests on Means after *ANOVA*

Please make certain you have attempted SAQ 1.4a and have read the responses before looking at this section.

The result of the *ANOVA* calculation in SAQ 1.4a showed clearly that there was a significant difference in the titration means, that is to say there is a high probability of a difference between the students. We now wish to know which students are significantly different on an individual basis. There are various procedures available for this such as the least significant difference (*LSD*), Scheffe's test, Tukey's paired comparison procedure, Newman–Keuls test and others. These tests require a measure of the residual error which is usually supplied by the residual error mean square. For this example we are going to use the least significant difference (*LSD*).

The individual students' averages for SAQ 1.4a were as follows:

Student	A	B	C	D
	25.06	24.91	25.14	25.17

First arrange the means in ascending order:

$$\bar{y}_B \qquad \bar{y}_A \qquad \bar{y}_C \qquad \bar{y}_D$$
$$24.91; \qquad 25.06; \qquad 25.14 \qquad 25.17.$$

In this method a quantity known as the *Standard Error of the Difference (sed)* is used to distinguish between the means.

$$sed = \sqrt{ResidMS} \times \sqrt{2/n}$$

The value n is the number of replicate titrations = 3, therefore

$$sed = \sqrt{0.00282} \times \sqrt{2/3} = 0.0531 \times 0.816$$

The test utilizes a reference t distribution with degrees of freedom given by the residual error mean square. If the differences between the means are greater than this value they can be said to be significantly different at that probability level. The critical value of t at $P = 0.05$ is found to be 2.31 from the t-tables.

The *LSD* value is given by

$$LSD = sed \times t_{resid\,d.f.} = 0.0531 \times 0.816 \times 2.31 = 0.10\ \text{ml}$$

where *resid d.f.* is number of degrees of freedom associated with this estimate (in this example, 8).

Comparing this *LSD* with the differences between the means suggests that there is no significant difference between $\bar{y}_C$ and $\bar{y}_D$, since $\bar{y}_D - \bar{y}_C = 25.17 - 25.14 = 0.03$ ml which is less than the *LSD* value (0.10 ml). However, $\bar{y}_B$ and $\bar{y}_A$ may be said to differ significantly from each other at the $P = 0.05$ probability level since $\bar{y}_B - \bar{y}_A = 25.06 - 24.91 = 0.15$ ml, which is greater than the 0.10 ml *LSD* value. Both $\bar{y}_B$ and $\bar{y}_A$ also differ significantly from $\bar{y}_C$ and $\bar{y}_D$. You should note that this method, although useful, is not as rigorous as other more complicated tests.

1.5. SUMMARY

In this part we have examined various techniques such as the *t*-test, *F*-test and *ANOVA*, which you may have come across previously and, as such, may only have served to refresh your memory. However, these analytical techniques go hand in hand with the following parts, in which aspects of experimental design will be introduced.

Learning Objectives

After reading the material in Part 1 you should now be able to:

- identify the types of error that can occur in an analytical method;
- perform *t*- and *F*-tests and use significance testing;
- use significance tests based on the range of a set of results;
- perform one-factor *ANOVA* calculations.

SAQS AND RESPONSES FOR PART ONE

SAQ 1.1a

The following results were obtained by two analysts using a new method for the determination of nickel in a standard reference alloy containing 6.25% nickel.

Analyst A	Analyst B
6.16	6.50
6.41	6.54
6.23	6.52
6.28	6.56

Mark the following statements with a cross ($\times$) if wrong, and a tick ($\sqrt{}$) if correct:

(*i*) The results obtained by analyst A are more precise than those of analyst B;

(*ii*) Since the results of analyst A fall on both sides of the true value, they could not be attributed to random error;

(*iii*) The range of results obtained by analyst A is 0.25;

(*iv*) It appears that analyst B's results are subject to a systematic error;

(*v*) The standard deviation of the results obtained by analyst A is $0.105_5\%$;

(*vi*) The variance of the results of analyst A is $\sqrt{0.105_5} = 0.324_8$

Response

The answers should be (i) ×; (ii) ×; (iii) ✓ ; (iv) ✓ ; (v) ✓ ; (vi) ×.

If you answered the questions correctly you have shown that you are conversant with some of the basic terms and concepts which we will use later on. Let us now go through each question in some detail. Even if you answered all the questions correctly, please read my comments.

(*i*) The statement is wrong.

Precision is often defined as the reproducibility of a set of results. The results of analyst B all lie between 6.50 and 6.56%. Thus, you could say these results are reproducible although they are all higher than the true value of 6.25%. These results are therefore precise, having small random errors, but are also inaccurate because there is a systematic error. The results of analyst A lie between 6.16 and 6.41% and are therefore far less precise than those of analyst B. The mean value $\bar{y}_A$, 6.27%, achieved by analyst A is very close to the true value μ of 6.25%. Hence you can say that the results of analyst A are imprecise since there are comparatively large random errors, but they are accurate because there is only a small systematic error. You should be aware of the modern definition of precision which makes a distinction between the terms reproducible and repeatable.

If analyst B had used the same experimental conditions, using the same equipment and performed the four replicate analyses in rapid succession, the results may be termed repeatable. If, on the other hand, the data had been obtained on successive days using different equipment, the results may be defined as being reproducible.

(*ii*) The statement is wrong.

You can see from the definition of random errors that the individual results of analyst A lie on both sides of the true value of 6.25%. Hence the results may be attributed to random errors, which in this example are quite appreciable.

(*iii*) The statement is correct.

The values obtained by analyst A lie between 6.16 and 6.41%. Range, a measure of the spread of the results, is defined as the difference between the largest and smallest values. So in this example the range is $6.41 - 6.16 = 0.25\%$. You should see that the larger the value of the range, the more imprecise is the set of results.

(*iv*) The statement is correct.

It was stated in (*i*) that analyst B's results are inaccurate, although they are precise. Systematic errors affect the accuracy, which is the closeness of the measured results to the true value. Hence it would be quite reasonable to say analyst B's results were subject to systematic error.

(*v*) The statement is correct.

The standard deviation s is the most commonly used measure of the spread of a set of results.

The standard deviation is defined as:

$$s = \sqrt{\frac{\Sigma(\bar{y} - y_i)^2}{N - 1}}$$

and can be calculated as follows:

Analyst A	$(\bar{y} - y_i)$	$(\bar{y} - y_i)^2$
$y_1 = 6.16$	-0.11	0.0121
$y_2 = 6.41$	$+0.14$	0.0196
$y_3 = 6.23$	-0.04	0.0016
$y_4 = 6.28$	0.01	0.0001
$\Sigma_y = 25.08$	$\Sigma(\bar{y} - y_i)^2 =$	0.0334

The mean, $\bar{y} = 25.08/4 = 6.27$

Therefore, $s = \sqrt{0.0334/3} = 0.105_5\%$.

You may of course have obtained the values of $\bar{y}$ and s using a calculator or you could have access to a computer program to work out these results.

Obviously if N is large a computer program is most useful! If you have a hand calculator always use the "σ_{N-1}" key for evaluating the standard deviation, unless the value of N is very large (21 or more), when there is little difference between the σ_N and the σ_{N-1} values.

(*vi*) The statement is wrong.

The variance V, is defined as the (standard deviation)^2. So in fact the correct value of V for the results of analyst A is $(0.105_5)^2 = 0.011$. It is not the square root of the standard deviation.

You will see later that variance is a most important term. It is possible to obtain the sum of variances, whereas standard deviations are not additive. For example if the sampling error (standard deviation) was 3% and the analytical error was 1% then the total error would be given by

$$\text{Total error} = \sqrt{(3^2 + 1^2)} = 3.16\%$$

The answer would not be 4%.

****

SAQ 1.2a	In a paired *t*-test comparing two methods, X and Y, in which twelve samples were analysed, a *t* value of 2.20 was calculated. From this information and from *t*-tables select with a circle the correct answer to each of the following questions:
	(*i*) How many degrees of freedom are there? (11 or 12 or 13).
	(*ii*) For a two-tailed test what is the value of *t* at the $P = 0.05$ level given $N = 12$? (1.78 or 1.80 or 2.18 or 2.20). ⟶

SAQ 1.2a
(cont.)

> (iii) Does the calculated value of t (2.20) lead us to the conclusion that method Y gives significantly higher results than method X at the $P = 0.05$ probability level? (No, or they are the same, or Yes).

Response

The correct answers are: (i) 11; (ii) 2.20; (iii) "Yes".

I hope you answered all of these correctly. If you did, you've shown you are able to use the t-tables and can compare the value of t from the table to the calculated value of t.

(i) The first question is concerned with the number of degrees of freedom (d.f.), sometimes given the symbol v, but you may also find that the symbol ϕ is used. In this question $d.f. = N - 1$, where N is the total number of replicate analyses; so $d.f. = 12 - 1 = 11$.

You may be puzzled by the concept of degrees of freedom so let's look at this example.

Imagine you are asked to choose any two integers between 1 and 10. You can select say 2 and 3, or 6 and 9, or a fairly large variety of integers. If, however, you are asked for two integers between 1 and 10 whose sum = 10 you could select 3 and 7, 4 and 6 etc., but in each case your second choice is governed by the first integer you chose. In other words you have lost one degree of freedom. If you are now asked to select two integers between 1 and 10 whose sum = 10 and product = 24, you can only choose 6 and 4. You have now lost two degrees of freedom because in reality you have two equations to satisfy.

In obtaining the standard deviation we have used $\bar{y}$ in the equation and have lost one degree of freedom, i.e. $N - 1$.

(*ii*) If you look at the *t*-tables for 11 *d.f.* you should find the following values.

P = 0.05 (5%); 0.025 (2.5%); 0.01 (1%); 0.005 (0.5%)
t = 1.80   2.20   2.72   3.11

Remember that for a two-tailed test you have to double the value of P to obtain the correct probability level. Therefore, you use the one-tailed P = 0.025 (2.5%) level to obtain a t value of 2.20, which is the corresponding two-tailed value at the P = 0.05 (5%) probability level.

(*iii*) I guess this was rather a sneaky trick! Did you spot the word "higher" in the question? This means that you should apply a one-tailed *t*-test, so for *d.f.* = 11, at P = 0.05 (5%), the value of t is 1.80. Therefore, the calculated value of t is greater than the tabulated value and you can say that method Y gives significantly higher results than method X at the P = 0.05 probability level.

****

SAQ 1.2b   A chemical manufacturer wishes to test a new route (B) for the synthesis of an organic compound. The original method of synthesis (A) has been used by the company for many years. Purely on the statistical evidence of the yields given below, determine whether it is sensible for the manufacturer to change to method B for the production of the organic compound.

	Number of results	Yield (%) of the organic	Standard deviation
Method A	5	72.3	0.36
Method B	6	72.9	0.33

Response

This is an example of the use of the t-test for the comparison of two means, $\bar{y}_A$ and $\bar{y}_B$. The first part of this problem was to calculate the pooled standard deviation s_p:

$$s_p = \sqrt{\frac{(N_A - 1)s_A^2 + (N_B) - s_B^2}{N_A + N_B - 2}}$$

$$s_p = \sqrt{\frac{4(0.36)^2 + 5(0.33)^2}{5 + 6 - 2}} = 0.344$$

The value of s_p that you have calculated should lie between 0.36 and 0.33. If you get a value outside this range, you have made a mistake in the calculation! Once you have calculated this value you should have used the following equation to obtain t:

$$t = \frac{\bar{y}_A - \bar{y}_B}{s_p \sqrt{(1/N_A + 1/N_B)}}$$

Therefore,

$$t = \frac{0.6}{0.344 \times \sqrt{(1/5 + 1/6)}} = 2.88_5$$

You should then have looked up t-tables for $N_A + N_B - 2$ degrees of freedom. At $d.f. = 9$ you will have found the following:

P	=	0.05	(5%);	0.025	(2.5%);	0.01	(1%);	0.005	(0.5%)
t	=	1.83		2.26		2.82		3.25	

A one-tailed test is used because it is reasonable to assume that the manufacturer wants to obtain a higher yield. The mean yield obtained by method B can thus be said to be significantly higher than the mean yield obtained by method A at the $P = 0.01$ probability level (because $2.88_5 > 2.82$).

Even if you had used a two-tailed test you could have said that the mean yield of method B is significantly different from that of method A at the 5% level but is not significant at the $P = 0.01$ level.

$$**********************************$$

SAQ 1.3a

> A new method for the determination of dissolved oxygen in water was compared with the standard (Winkler) method. The following results were obtained for a river water sample.
>
> New method: 62; 66; 63; 65% dissolved O_2
> Standard method: 67; 66; 66; 67% dissolved O_2
>
> Use the F-test and the F_R test to examine whether the precisions of the methods differ significantly?

Response

First you had to work out the variances of each set of results. The variance V can be obtained from the formula

$$V = \Sigma(\bar{y} - y_i)^2/(N - 1)$$

You will obviously have shown that the mean of the first set of results $\bar{y}_1 = 64$, and the mean of the standard method $\bar{y}_2 = 66.5$.

I calculated the variance V_1 as follows:

$\Sigma(\bar{y} - y_i)^2 = 10$ and therefore $V_1 = 10/3 = 3.33$ and similarly $V_2 = 0.333$. I then calculated an F-ratio:

$$F = V_1/V_2 = 10$$

From the F-tables for 3 degrees of freedom for each set of values at the $P = 0.05$ probability level, $F = 9.28$. Since $10 > 9.28$, you should have concluded that there is a significant difference in the precisions of the two methods. (If you had difficulty in getting the value of F from the tables, please refer to Fig. 1.3a).

Using the test based on the ranges (W) of the results, $W_1 = 66 - 62 = 4$ and $W_2 = 66 - 65 = 1$.

Hence,

$$F_R = W_1/W_2 = 4.0$$

I hope you agree that these calculations are very easy! The critical value for F_R at $P = 0.05$ for four pairs of values $= 4.0$, so the precisions of the two sets of results are significantly different at exactly the $P = 0.05$ level. You should realise that it is necessary to have exactly the same number of results in each set of data to perform the F_R test. Also you do not have to concern yourself with degrees of freedom in this case because the tests use only the ranges and the mean values are not used.

****

SAQ 1.4a	Four students were each asked to perform three replicate titrations using the same titrimetric procedure. The end-points obtained by each student are shown in the following table.

Student	A	B	C	D
	25.03	24.95	25.13	25.16
Titres	25.09	24.90	25.20	25.25
(ml)	25.06	24.89	25.08	25.10

$\longrightarrow$

SAQ 1.4a
(cont.)

Determine whether there is a significant difference (at the $P = 0.05$ level) in the mean titration value obtained by each student. Present your final results in an *ANOVA* table.

Response

This is a one-way analysis of variance calculation.

Let us go through the calculation step by step and I will make some comments at various stages on the solution to this problem. The data is as follows:

Student			
A	B	C	D
25.03	24.95	25.13	25.16
25.09	24.90	25.20	25.25
25.06	24.89	25.08	25.10

I hope you remembered that it is good practice to subtract a common quantity from each value. It makes the calculation much easier to perform. I have decided to subtract 25.00 from each value. You may have selected a different number, but provided you have not made an error in the calculation you should obtain exactly the same values in the final *ANOVA* table. So, by subtracting 25.00, I obtained the following values:

	A	B	C	D
	0.03	−0.05	0.13	0.16
	0.09	−0.10	0.20	0.25
	0.06	−0.11	0.08	0.10
Sum	0.18	−0.26	0.41	0.51

I then obtained the *Grand Total* (T) from

$$0.18 - 0.26 + 0.41 + 0.51 = 0.84$$

The correction factor $CF = T^2/N = 0.84^2/12 = 0.0588$

Please note that I have not rounded off the value of CF. The next step is to calculate the total corrected sum of squares.

I had to square each individual reading and then sum the columns to obtain the crude sum of squares $(CrudeSS)$:

	A	B	C	D
	0.0009	0.0025	0.0169	0.0256
	0.0081	0.0100	0.0400	0.0625
	0.0036	0.0121	0.0064	0.0100
Sum	0.0126	0.0246	0.0633	0.0981

I then added these column totals together;

$$0.0126 \quad + \quad 0.0246 \quad + \quad 0.0633 \quad + \quad 0.0981$$

$$= 0.1986$$

The *TotalSS* = *CrudeSS* − *CF* = 0.1986 − 0.0588 = 0.1398

Remember, the total corrected sum of squares must be positive if your value is negative, then you have made a mistake. The next stage is to determine the between-student or between-treatment sum of squares. I then squared the sum totals of each column and divided this by the number of readings in each column (3) and subtracted the CF. Thus,

$$TreatmentSS = \frac{(0.18)^2 + (-0.26)^2 + (0.41)^2 + (0.51)^2}{3} - 0.0588$$

$$= \frac{0.0324 + 0.0676 + 0.1681 + 0.2601}{3} - 0.0588$$

$$= 0.11727$$

Please note this value must also be positive and must be less than the total corrected sum of squares. The most common mistake is to divide by the number of columns rather than the number of individual results in each column. The residual error sum of squares is just the *TreatmentSS* subtracted from the *TotalSS*. Therefore, the residual error sum of squares = $0.1398 - 0.11727 = 0.02253$.

The number of degrees of freedom (*d.f.*) is as follows:

$$Total\ d.f. = (N - 1) = 11$$
$$Treatment\ d.f. = (t - 1) = 3$$

where t = the number of columns.

The residual error $d.f. = N - t = 8$. I can now set up the *ANOVA* table, and you are now in the position to check the final values if you subtracted a different amount at the start of this calculation. If you divide the sum of squares values by these *d.f.* values the mean squares values are obtained.

Source of variation	d.f.	SS	MS
Between-students	3	0.11727	0.03909
Residual error	8	0.02253	0.002816
Total	11	0.1398	

The F ratio = $0.0391/0.00282 = 13.86$

Referring to the F-tables for three and eight degrees of freedom at the P = 0.05 probability level, I found that $F = 4.07$. Since the calculated value of F (13.86) is greater than this, it is apparent that the titration means differ significantly. I hope you managed to get the right answer and, just as importantly, arrived at the correct conclusions from the F-test. After congratulating yourself, please read on. If you look again at the question you should see that, although there is a difference in the titration values between the students, there is little difference in the precision of each student. The within-student or residual error sum of squares is an indication of the error variance, a measure of the random error obtained with replicate titrations. After you have realised there is a significant difference in the titration means you should find out which means differ on an individual basis. Please refer to Section 1.4.4 where this problem is discussed.

****

2 Principles of Experimental Design

2.1. INTRODUCTION

Let us consider the Kjeldahl analysis of soils for total nitrogen. You may or may not be familiar with this technique, but in either case I shall point out that the final determination by titration with dilute HCl is usually preceded by a hot acid digestion which is promoted by a catalyst, and steam distillation after addition of conc. NaOH to remove the ammonium-nitrogen from the digestate. The concentration values determined by the analyst could be influenced by, amongst others, the following experimental variables:

> the acid or mixture of acids used;
> the catalyst used;
> the temperature of the digestion;
> the time of the digestion.

Suppose we are going to attempt this analysis for the very first time on an unusual type of soil. How do we go about it? Do we use conditions which other workers have used? It might be a useful start, but, if we have some problems with the technique, do we then attempt to obtain a "good response" by the often suggested "trial and error" method or should we try to plan our approach prior to experimentation? It is invariably true that we should adopt the latter course of action. Often we waste time and effort by rushing into the experimental work and, most importantly, we frequently do not have any idea of the significance of the results we have obtained. So how do we go about planning the experiment(s)?

Π   In the example quoted above we identified four variables; can you
       think of other variables that would affect the response?

Well, you may have considered the amount of the catalyst used, which
should be sufficient to promote the conversion of the organic forms of
nitrogen to ammonium-nitrogen in the time set aside for the digestion. Also
we have not mentioned variables which influence the steam distillation.
How much conc. NaOH should we add? Also, how long should we carry
out the distillation for? This is not a complete list and you might have
thought of others.

The important point is that you have stopped to think about these factors
prior to performing the experiment. This is the first stage of what is called
experimental design. You can see that I have introduced the term *factor*,
which may be defined as any experimental variable that can affect the
result of an experiment. The result or value obtained from an experiment
is called the *response*. The response in this case is the concentration of
nitrogen but it could be an absorbance reading in spectroscopy, or a more
complicated value like the chromatographic response function which takes
into account both peak separation and peak retention times. Thus in an
experimental design we must be clear exactly what we wish to obtain from
the experiment.

Having identified the factors that may affect the response, we now have to
determine the *levels* of some factors. In our Kjeldahl digestion experiment
we may wish to investigate the effect of changing the digestion temperature
from 230 °C to 260 °C. The different values a factor takes are referred to
as different levels. Thus the two levels of temperature are 230 °C and
260 °C respectively. Some factors are termed *qualitative* in respect of the
fact that they may only be present or absent, on or off and so on. Other types
of factors are *quantitative*, such as concentration, which may take any one
of a number of levels. Once we have determined the factors and decided
what levels they should take we have to decide upon the combination of
levels of the factors and how many runs are going to be carried out.

The conditions which are particular to one experiment are termed a
treatment, and we could select a treatment with a reaction temperature
of 260 °C and an allowed digestion time of 120 minutes. Another treatment

might well have the same digestion temperature but have an increased digestion time of 180 minutes. Variables such as temperature are called controlled factors, since we can select the reaction temperature we desire. Other factors may be out of our control. For example, there may be different workers performing the final titrations. The techniques of these workers may vary. This variable would be defined as an uncontrolled factor. The second stage of designing an experiment is to attempt to minimize the effects of uncontrolled factors, and we shall see in the next section that we can use the technique of *randomization* to take into account the effect of uncontrolled factors. Finally we carry out the experiments, analyse the results using one or more of the statistical methods available, and if necessary carry out any transformations of the data and re-analyse the results.

You should realise that the examples we are discussing are in some ways unimportant and that they are only introduced to put over particular points about experimental design. You may find some of the examples we use to be irrelevant to you, and indeed this may even be true for all the examples, but if you understand the concepts you will have derived some benefit from them and can take the ideas and apply them to your experimental situations.

2.1.1. Replication

Suppose we now decided to carry out two experiments, one at each of two temperatures, keeping all other factors constant, i.e. two treatments. If we obtain a single value of concentration from each treatment we have no way of estimating the residual error (introduced in Part 1). Also we increase the risk of drawing conclusions from what may be spurious results. Therefore, we must *replicate*, that is obtain repeated measurements of nitrogen concentration for each treatment. We could do this by using two replicates for each temperature. In such a case the number of replicates is deliberately *balanced* between temperatures. Balance is an important property of experimental design which we will return to in the next section.

2.1.2. Randomization

Suppose we decide to examine the effects of four temperatures with four replicates for each temperature, over the course of one day. We could apply the temperatures so that the first four experimental runs receive the lowest temperature (A), runs 5–8 receive the next highest temperature (B) and so on. However, unknown to us an uncontrolled factor causes higher results at the early part of the day and decreases during the day. We are therefore running the risk of obtaining biased results, with temperature A treatments being measured at times with naturally high values and temperature C and D treatments at times of low values. Comparison of the effects of temperature would be unfair because the concentrations measured at those temperatures are unrepresentative.

∏   Would a design with the order ABCD in the first block, ABCD in the second block etc. minimize the tendency of obtaining biased results?

No it wouldn't, because there could be other trends throughout the day which are not as obvious. For example, whoever is working on the apparatus might leave it switched off after every four runs while he or she went for a break or lunch. Suppose that a stirring apparatus is employed in the critical final titration stage and that this apparatus takes some time to warm up and function at its set speed. The end-point of this titration is often determined by monitoring the pH, and a variable stirring rate could have important effects upon the values obtained. The runs at the beginning of each session might then have lower values than those at the end of a session, so biasing the results. Any deliberate pattern of applying the four temperatures might thus be subject to bias from trends throughout the day.

Randomization is a way of guarding against such extraneous variation whether or not we are aware of it. Instead of applying the treatments in one of the orders I mentioned above, we simply apply them in a random order. There are a number of ways of accomplishing this. We could stick a pin (without looking) into a page of random numbers and look down the page for numbers 1 to 16 until all have been selected. Alternatively, you could have access to a piece of software which will either generate random numbers between 1 and 16 or, if it is software written with experimental design in mind, it could randomize the order of the runs for you. Probably the simplest method is to write the 16 letters on pieces of paper and draw them from a hat or envelope. However, it is possible (though unlikely) that you could end up with a pattern with all A treatments first and so on. Assigning the treatments in a random order does not therefore completely remove the possibility of biased results, but does reduce the chances.

2.2. RUNNING EXPERIMENTS IN BLOCKS

One of the assumptions of *ANOVA*, which is used to analyse experimental results, is that the uncontrolled variation is random. This random variation is used to test whether the treatments are different, and replication is usually used as the source of such variation. Sometimes, however, it is not possible to carry out all the experimental runs under exactly the same conditions. For example, the procedure may be lengthy and several days may be required to carry out all the experimental runs. The runs on different days may therefore be conducted under different conditions which could affect the values obtained. Alternatively, there may be insufficient raw material for all the runs and more than one *batch* (termed *blocks*) of material has to be used. Observations within blocks can therefore be compared with greater precision than observations distributed over the whole experimental material. As a result the *randomized block design* has been developed. With a randomized block design the resulting *ANOVA* can separate the variation due to the treatments, blocks and residual error.

Suppose in the above example the effects of the four temperatures (A, B, C and D) upon the concentration of nitrogen are to be investigated but, because of the amount of soil which is available at any one time, only eight runs may be carried out with any one batch. One possible design is

    Batch 1   A   A   C   A   D   A   C   D
    Batch 2   C   B   C   D   B   B   D   B

If B turns out to be significantly different from A it will not be possible to say whether this is due to an actual difference between the treatments or because of the differing soils.

∏   Can you think of a design for these four temperatures in two batches in which the effects of the treatments can be separated from the effects of the batches?

This is a tricky one because you have to realise that each temperature should be tested twice in each batch, with the order of the runs in each batch being randomized in an attempt to avoid any time trends influencing the results. One that I came up with is given below. Of course there is only a small chance that your random order is the same as mine but the important point is that the treatments are tested twice in each batch.

Batch 1   A   B   D   C   C   A   B   D

Batch 2   C   B   A   A   D   B   D   C

Any differences between the effects of A and B will now not be completely confused with possible differences between the batches. As with most designs, information on the effects of blocking and the factor is most easily obtained if the randomized block design is balanced so that the treatments are examined an equal number of times in each block. This property ensures that all the effects can be estimated directly and independently of one another, a condition known as *orthogonality*. The sums of squares of all the effects are then additive. If something went wrong with one of the batches so that only the results of three temperatures replicated twice in each batch were obtained, the average effect for the remaining temperatures will be influenced because it has only been obtained over one of the batches. This has an effect upon the resulting analysis of variance because differences between the averages are used.

In blocked designs the effects of the factors are of interest whereas the effect of the block is usually of no interest and is to be eliminated. Also in using a blocked design you are assuming that the effects of the factors are the same over all the blocks. If this is not true there is said to be an interaction between the blocks and the factors. I will return to this in later parts.

SAQ 2.2a

> Suppose that you have to examine the effects of three treatments (we shall call them A, B, C) with four replicates in twelve runs. Unfortunately the experimental procedure is lengthy and you can only manage three runs per day. Produce a randomized block design which will enable you to decide what effects the treatments have, whilst at the same time estimating the residual error and block (day-to-day) effect.

The responses to the SAQs in Part 2 begin on page 73.

2.2.1. Worked Example Randomized Block Design and *ANOVA*

We have decided to carry out the experiment on the determination of nitrogen in soils by Kjeldahl digestion but, instead of the four or five factors identified previously, we only examine the effect of one factor, namely temperature. We have selected four temperatures, 230, 260, 290 and

320 °C. However, as a result of time and cost limitations it is only possible to have sixteen experiments, and therefore a maximum of four replicates for each temperature. Due to the scale of the operation there is insufficient soil for all sixteen experimental runs and we use a randomized block design for the two soils, carrying out the runs in the random order given in Section 2.2. The results are given below:

Treatment	N_2 Concentration Soil 1		Soil 2		Average N_2 Conc. (%)
A 230 °C	8.2,	8.4	8.2,	8.4	8.3
B 260 °C	8.6,	8.8	8.4,	8.2	8.5
C 290 °C	7.8,	8.0	7.8,	7.6	7.8
D 320 °C	7.0,	7.3	7.8,	7.5	7.4

Obviously the mean treatment values are different, but because of random error and possible differences between the soils, even if the true value is unchanged, the treatment means may vary from one treatment to another. If we adopt a null hypothesis that the treatments (temperatures) and batches (soils) have no effect upon the response, we can estimate the variance in the data in three ways; one derived from the variation between the treatments, another from between the soils and finally the residual variance.

The overall average (8.0%) has been subtracted from each response to simplify the calculations.

Treatment	N_2 Conc. % Soil 1		Soil 2		Average N_2 Conc. (%)
A 230 °C	0.2,	0.4	0.2,	0.4	0.3
B 260 °C	0.6,	0.8	0.4,	0.2	0.5
C 290 °C	−0.2,	0.0	−0.2,	−0.4	−0.2
D 320 °C	−1.0,	−0.7	−0.2,	−0.5	−0.6

We can now calculate the sums of squares quantities as we have done previously. Here we will give the between-treatment sum of squares the abbreviation *TreatmentSS* and the between-soil batch sum of squares is abbreviated to *BatchSS*.

∏ From what you have learned about the additivity of the sum of squares terms in the previous part, write down an expression for the *TotalSS* in terms of the *TreatmentSS*, *BatchSS* and *ResidSS*.

I hope you remembered from Part 1 that sums of squares terms are additive, so it is possible to write a simple equation expressing this for the blocked experiment:

$$TotalSS \; = \; TreatmentSS \; + \; BatchSS \; + \; ResidSS$$

In single-factor *ANOVA* with the same number of replicates to each treatment we defined the total number of experiments as

$$N \; = \; r \, \times \, t$$

where r is the number of replicates and t is the number of treatments. With a blocked design we have to include the number of batches (b), so that

$$N \; = \; b \, \times \, r \, \times \, t$$

where b is the number of (soil) batches and r is now the number of times a treatment is used within a soil batch. We now need to work out these sum of squares quantities. However, some of the formulae we used earlier have to be adjusted to incorporate batches. First of all we need to define a response in terms of the soil from which it is taken. We will now give the response the symbol

$$y_{ijk}$$

where k is the replicate within the soil (varying between 1 and r, or r_t when the number of replicates is not the same for every treatment), j is the treatment (varying between 1 and t), and i is the batch (varying between 1 and b)

Π   In the above example what is the response of y_{231}?

The answer you should have given is 7.8% since the first subscript (2) refers to the second soil, the second subscript (3) refers to the third treatment and finally the third subscript (1) refers to the first replicate. You should have looked this position up in the table of original responses to find 7.8%.

In Part 1 we defined the between-treatment sum of squares as the sum of the squares of the treatment totals divided by the number of replicates, with a final subtraction of the correction for the mean. We now have to include soil batches in this calculation so that:

$$TreatmentSS = \frac{\sum\limits_{j=1}^{j=t} \left(\sum\limits_{i=1}^{i=b} \sum\limits_{k=1}^{k=r} y_{ijk} \right)^2}{r \times b} - T^2/N$$

Similarly, to calculate the *BatchSS* we sum the responses within each soil batch, square them, sum these squares for all the soils and divide by the number of replicates (r) multiplied by the number of treatments and finally subtract the correction for the mean so that:

$$BatchSS = \frac{\sum\limits_{i=1}^{i=b} \left(\sum\limits_{j=1}^{j=t} \sum\limits_{k=1}^{k=r} y_{ijk} \right)^2}{r \times t} - T^2/N$$

Probably the best way to work out these values is to set up a two-way table as given below:

Batch		Treatments			Batch total	$\left(\dfrac{Batch}{total}\right)^2$	$\Sigma\left(\dfrac{Batch}{total}\right)^2$
	A	B	C	D			
Batch 1	0.2	0.6	−0.2	−1.0			
	0.4	0.8	0.0	−0.7	0.1	0.01	0.02
Batch 2	0.2	0.2	−0.2	−0.5			
	0.4	0.4	−0.4	−0.2	−0.1	0.01	
						0 = *Grand total (T)*	
Treatment total	1.2	2.0	−0.8	−2.4			
$\left(\dfrac{Treatment}{total}\right)^2$	1.44	4.00	0.64	5.76			
$\Sigma\left(\dfrac{Treatment}{total}\right)^2$ 11.84							

Therefore, $TreatmentSS = 11.84/4 = 2.96$

and $BatchSS = 0.02/8 = 0.0025$

We can calculate the total corrected sum of squares as before by summing the squares of the individual values in the above table. Please note that it is now unnecessary to subtract a correction for the mean term as this is zero ($0.0^2/16 = 0$).

$$TotalSS = 0.2^2 + 0.4^2 + 0.2^2 + \ldots + (-0.5)^2 + (-0.2)^2$$
$$= 3.62$$

ResidSS can be calculated by subtraction:

$$ResidSS = TotalSS - TreatmentSS - BatchSS$$

Therefore $ResidSS = 3.62 - 2.96 - 0.0025 = 0.6575$

We can now set up an analysis of variance table:

Source of variation	d.f.	SS	MS	Variance ratio
Treatment	3	2.96	0.9867	16.51
Batch	1	0.0025	0.0025	0.04
Residual	11	0.6575	0.05977	
Total	15	3.62		

The three between-treatment *d.f.* were derived from the number of treatments minus one. The between-batch *d.f.* were derived from the number of soil batches minus one (2 − 1 = 1) and the residual error *d.f.* from the total *d.f.* minus the between-treatment *d.f.* and between-batch *d.f.* (15 − 3 − 1 = 11). Here again we have already subtracted the single *d.f.* associated with the average. The mean square (*MS*) terms in the table were obtained by dividing the sums of squares terms by their degrees of freedom. These mean square terms now give us three estimates of the variance. If the null hypothesis is correct then these estimates should be the same. The variance ratios enable us to decide whether this is likely at a given probability level.

The between-treatment to residual error variance ratio is 16.51, much greater than the tabulated value of 3.59 for a one-tailed *F* distribution with 3 and 11 *d.f.* at a $P = 0.05$ probability level. We can therefore reject the null hypothesis that the treatments are the same and accept the alternative hypothesis that they are different.

The variance ratio for the batches (0.04) is obviously less than the tabulated *F* value of 4.84 for 1 and 11 *d.f.* ($P = 0.05$). Therefore, we may not say that variation between the batches has had a significant effect upon % yield at a probability level of $P = 0.05$. However, we still do not actually accept the null hypothesis since there is always the possibility that if we repeated the experiment we could actually find a significant difference.

SAQ 2.2b

In an experiment examining the effect of six batches of material and four treatments in 48 experiments, how many degrees of freedom can be allocated to each of the sum of squares values?

SAQ 2.2c

In an experiment to compare the percentage efficiency of four different chelating resins in extracting Cu^{2+} ions from aqueous solution, the experimenter decided to use a randomized block design since it was only possible to carry out four runs with each resin. On each of three successive days a fresh solution of the Cu^{2+} ions was prepared and the extraction was performed with each of the chelating resins, taken in a random order.

A randomized block design was set up and the following results were obtained:

	Chelating resin			
Day	A	B	C	D
1	97	93	96	92
2	90	92	95	90
3	96	91	93	91
4	95	93	94	90

Decide by an appropriate analysis whether the differences between the resins and the days on which the analyses were conducted had significant effects at the $P = 0.05$ probability level.

2.2.2. Balanced Incomplete Block Designs

When it is not possible to run all the treatments within one block, so that all the treatments cannot be tested under uniform conditions, the experimenter has to find a way of running the blocks without affecting the comparisons between the treatments. Such a situation might occur in chemical experimentation when the available batches of raw material of uniform quality are not sufficient for more than a small number of experiments. When all comparisons are of equal importance, the treatments should be selected in a balanced way, so that any two occur together the same number of times as any other two. Such designs are termed *balanced incomplete block designs*. In some ways these designs are like having missing results from the randomized block designs discussed in the previous section. However, when results are removed in an ordered way the analysis is easier (though still not very easy) than if results are lost at random (when the design is an *unbalanced incomplete block design*).

The simplest balanced incomplete block design is given below and was set up by an experimenter who wanted to use three treatments but could only run the experiments in blocks of two observations (y).

Treatment	Block			
	1	2	3	
				replicates $r = 2$
A	y		y	treatments $t = 3$
B	y	y		blocks $b = 3$
C		y	y	block size $m = 2$

Note that here the term replicates (r) is used as the total number of times a treatment is tested, but not necessarily under the same conditions.

∏   How many times does a given treatment appear with any other treatment?

Once. To find this out just look for any two treatments, say A and B, which only occur together in block 1 but in no other blocks. An easy method of working this out is given below;

$$\tau = r(m - 1)/(t - 1)$$

where τ (tau) is the number of blocks in which any two treatments occur together. Thus in the above example

$$\tau = 2 \times (2 - 1)/(3 - 1) = 1$$

These incomplete block designs can be symmetrical when the number of treatments is equal to the number of blocks. It is then possible to calculate the block effects corrected for differences between treatments.

∏   Does the following design represent a balanced incomplete block design?

Treatment	Block						
	1	2	3	4	5	6	7
A		y	y	y			y
B	y	y		y		y	
C	y		y			y	y
D		y	y		y	y	
E				y	y	y	y
F	y		y	y	y		
G	y	y				y	y

Yes, because each pair of treatments occurs twice. For example, treatments A and B occur together in the blocks 2 and 4 but in no other blocks, treatments B and C only occur together in the blocks 1 and 6 and so on.

∏   Is the design symmetrical?

Yes, because the number of treatments is equal to the number of blocks.

It is possible for you to work out these designs for however many blocks and treatments you have available by writing down all the possible combinations of the treatments. However, these *combinatoric* designs tend to require a larger number of runs and are not always efficient in this respect. Therefore, if you are in need of a balanced incomplete block design I would advise you to look it up in one of the general texts cited in this book.

Randomization of these designs is as necessary as with any other blocked design. There are several steps to this with balanced incomplete blocked designs. First of all you can randomize the order of the blocks, then randomly allocate the treatments to the letters (or whatever you are using) and finally randomize the positions of the treatments within the blocks.

Obviously if you do not do this, and carry on using the same design repeatedly, you are running the risk of introducing a bias, particularly when there is a possibility of consecutive observations being correlated.

Although these designs are extremely useful in certain circumstances their analysis is not as straightforward as for the complete block designs. It will therefore not be covered here as it has been extensively covered in some of the general textbooks on design and analysis of experiments.

2.3. LATIN SQUARES

Blocked experiments have been introduced to avoid the possibility of introducing bias where there are some inhomogenieties in the experimental material or where changes in the response with time can influence the results. However, let us now suppose that, as well as differences in batches of material, some changes in the response can be expected from the time of day at which an analysis or experiment is carried out, that is to say we effectively have two blocking variables. What designs are now applicable? Given below is a simple randomized block experiment for four blocks (1, 2, 3 and 4) and four treatments (A, B, C and D) conducted at four specific times of the day;

Block	Order during day			
	I	II	III	IV
1	A	C	B	D
2	A	D	C	B
3	C	A	D	B
4	B	A	C	D

Despite randomization treatment A is still run first or second within each block and is still susceptible to an influence of the change throughout the block. Let us suppose that the block is actually the day and that the experimenter can only run the experiments one at a time. If, for example, the experiment is an analysis in which light susceptible reagents are used which have to be made up fresh each day, it is possible that they could go off to some extent during the day. An analysis conducted in this manner could be severely affected. Treatment A might be found to have an average response somewhere near its true value whereas the other treatments would tend to have comparatively low average responses. It is necessary to have designs which can take two blocking variables.

Latin squares are a class of design which can be used to control the variation of a second blocking variable. Given below is one possible 4 × 4 latin square with four treatments (A, B, C and D):

Day	Time of day			
	I	II	III	IV
1	A	B	C	D
2	C	D	A	B
3	B	C	D	A
4	D	A	B	C

Each treatment now occupies each row and column of the design once only. This design allows the subsequent analysis of variance to separate the between-treatment, between-day, between-time of day and residual error variations.

∏   Design a 3 × 3 latin square for two blocking variables (days and time of day) and three treatments (A, B and C).

Fortunately there are only two possible 3 × 3 latin square designs. Using letters for treatments, which can of course be allocated randomly, both the following are correct:

Day	Time of day			Day	Time of day		
	1	2	3		1	2	3
1	A	B	C	1	A	B	C
2	B	C	A	2	C	A	B
3	C	A	B	3	B	C	A

Here again the rows, columns and treatments have to be randomized to avoid trends within the design.

∏ Can you spot one fairly obvious disadvantage to the latin square design?

From what you have seen of the above latin squares you may have observed that these designs must have the same number of treatments as levels of both blocking variables. If you are unable to divide the material or time (or other blocking variable) into equal units for both blocking variables, other types of design might have to be used. Despite this disadvantage these designs have been applied very frequently in industrial situations.

Let us suppose now that the chelating resin example used in SAQ 2.2c was in fact a 4 × 4 latin square in which the *Time of day* was used as another blocking variable in addition to *Day* and the experiments were carried out according to the 4 × 4 latin square design given above;

Day	Time of day			
	I	II	III	IV
1	A 97	B 93	C 96	D 92
2	C 95	D 90	A 90	B 92
3	B 91	C 93	D 91	A 96
4	D 90	A 95	B 93	C 94

In order to show you how to work out the analysis of a latin square design I have again subtracted the average (93%) from each response to get rid of the correction for the mean term. The results are shown in the following table which now has row and column totals assigned to work out the two blocking variable sum of squares.

Day	Time of day				Day total	Treatment total	
	I	II	III	IV			
1	A 4	B 0	C 3	D −1	6	A	6
2	C 2	D −3	A −3	B −1	−5	B	−3
3	B −2	C 0	D −2	A 3	−1	C	6
4	D −3	A 2	B 0	C 1	0	D	−9

Time of day
total   1   −1   −2   2

As before we have to square these values, sum them, and divide by the number of values contributing to them (four in each case). Therefore,

$$Between\ Time\ of\ daySS = [1^2 + (-1)^2 + (-2)^2 + 2^2]/4 = 10/4 = 2/5$$

$$Between\ DaySS = [6^2 + (-5)^2 + (-1)^2 + 0^2]/4 = 62/4 = 15.5$$

$$Between\ TreatmentSS = [6^2 + (-3)^2 + 6^2 + (-9)^2]/4 = 162/4 = 40.5$$

If you take a look at the response to SAQ 2.2c you will see that the sum of squares terms for between-days and between-treatments take exactly the same values in the randomized block and latin square designs.

∏   What is now the value of the residual sum of squares?

Since the sums of squares terms always sum to the total corrected sum of squares, already calculated (80), the residual sum of squares can be worked out by difference. Therefore,

$$ResidSS = 80 - 40.5 - 15.5 - 2.5 = 21.5$$

∏ How many degrees of freedom are associated with the between-time of day and residual sum of squares?

Three for between-time of day, since we used four different times of the day as the second blocking variable. Of course this leaves us with fewer degrees of freedom available for the residual error which we find by difference;

$$Resid \; d.f. = \underset{\text{(total)}}{15} - \underset{\text{(treatments)}}{3} - \underset{\text{(days)}}{3} - \underset{\text{(time of day)}}{3}$$

$$= 6$$

We are now able to form a new *ANOVA* table:

Source of variation	d.f.	SS	MS	Variance ratio
Treatment	3	40.5	13.5	3.768
Days	3	15.5	5.167	1.442
Time-of-day	3	2.5	0.833	0.232
Residual	6	21.5	3.583	
Total	15	80		

Unfortunately, the critical value of F at a $P = 0.05$ probability level for three and six $d.f.$ is 4.76, which is greater than all the calculated variance ratios. As a result we are unable to state that these blocking variables and treatments had any effect upon the amount of Cu^{2+} ions recovered by the resins. This is in direct contrast to the randomized block design with the same responses, in which treatments were significant at the same probability level.

This often occurs with designs in which there are relatively few degrees of freedom available for the residual error sum of squares. In the randomized block design the variance ratios were formed with nine $d.f.$ in the denominator, whereas in the latin square design only six $d.f.$ were available (three lost to the second blocking variable). Consequently the tabulated value which had to be exceeded to reject the various null hypotheses was higher.

When you are able to include a degree of replication in the latin square design, you may do this in a number of ways, each of which influences the subsequent analysis. For example, the latin square may replicated using the same blocking variables (this is the form you will meet most often), or using different versions of one or two of the blocking variables. A replicated 4 × 4 latin square requires a total of 32 experiments and hence has 31 *d.f.* available to the sum of squares terms, 16 of which are associated with the between-replicates sum of squares and may be used in the residual error estimation.

I am not going to run through more *ANOVA* at this stage as they tend to be a bit tedious to work out once you have done a few and, besides, you might concentrate too much on these aspects and miss some of the important aspects of experimental design. Anyway, what you have already learned about *ANOVA* should stand you in good stead to attack a real statistics textbook!

2.3.1. Graeco-Latin Square Designs

If you have a third blocking variable it may be incorporated in a *Graeco-Latin Square*. For example, to form a 3 × 3 graeco-latin square you have to superimpose the two 3 × 3 latin square designs;

A	B	C		A	B	C
B	C	A		C	A	B
C	A	B		B	C	A

but now substituting greek letters for the third blocking variable which are taken from the second design (hence graeco-latin). Therefore, the 3 × 3 graeco-latin square is

$$A\alpha \quad B\beta \quad C\gamma$$
$$B\gamma \quad C\alpha \quad A\beta$$
$$C\beta \quad A\gamma \quad B\alpha$$

Again these designs are most effective if they are replicated, otherwise you will use up all the available degrees of freedom. They are also subject to the same rules of randomization which have been put forward for the latin squares.

∏   Can you guess now what a *Hyper-Graeco-Latin Square design* might be and why a 3 × 3 hyper-graeco-latin square design is not possible?

If you have given the answer that it is a design with even more blocking variables than the graeco-latin square you are really thinking along the right lines as far as blocking is concerned. You should also have quickly realized that a 3 × 3 hyper-graeco-latin square is impossible because there are only two possible variants of the 3 × 3 latin square which can be incorporated to form the graeco-latin square design. The smallest hyper-graeco-latin square is a 4 × 4, formed by superimposing the only three possible 4 × 4 latin squares. Graeco-latin square designs can be formed by superimposing any two of these three. You have to be quite careful when trying to find appropriate graeco-latin square designs since some do not exist (e.g. 6 × 6). Also, you are best advised to look them up rather than work them out since this can take some considerable time.

2.3.2. Youden Squares

Just as balanced incomplete blocks can be used when the size of the block is too small to accommodate all the treatments, *Youden squares* have been devised as a means of running experiments with more than one blocking variable with incomplete blocks, that is to say, Youden squares are to latin squares as balanced incomplete block designs are to randomized block designs. These again have found much favour in industrial situations but have been relatively little used in analytical situations. Like balanced incomplete block designs their analysis is complex and I will leave it to you to look up an example if you think you have a use for such a design.

SAQ 2.3

Complete the following paragraph by inserting the most appropriate word or phrase, chosen from the list given below, into the blank spaces.

By an experimental design we mean a planned series of operations called .... in which certain conditions are held as constant as possible throughout and others are varied under the close control of the experimenter. The conditions which are particular to one experiment comprise the .... When more than one variable is changed between experiments the individual conditions are called .... The particular values at which an experiment are run are called the factor .... and where identical treatments are possible .... is used. If the experimenter is unable to keep certain conditions for one factor constant over the whole series of experiments .... is used to make comparisons more reasonable. A simple design for such a situation is the .... design. Sometimes, however, the blocks are of insufficient size to accommodate all the treatments and the experimenter may have to resort to .... designs. When two blocking variables are unavoidable a .... design may be used and when three blocking variables are necessary a .... design can be used. .... is a technique which must be applied to all these designs as a means of avoiding invalid conclusions. However, these designs suffer from the disadvantage that many of the available degrees of freedom are used up in estimating the block sum of squares. For example a 4×4 graeco-latin square design has only .... $d.f.$ available for the residual sum of squares.

$\longrightarrow$

SAQ 2.3
(cont.)

BLOCKING	FIVE	GRAECO-LATIN SQUARE
SIX	BALANCED INCOMPLETE BLOCK	TREATMENT
RANDOMIZED BLOCK	LEVELS	EXPERIMENTS
LATIN SQUARE	BALANCED COMPLETE BLOCK	
FACTORS SEVEN	REPLICATION	RANDOMIZATION

Learning Objectives

After reading the material in Part 2 you should be able to

- understand the elementary principles of experimental design such as randomization and replication;

- perform experiments in randomized blocks when insufficient material or time is available for the complete experiment;

- carry out *ANOVA* on randomized block designs;

- understand how to apply latin and graeco-latin square designs when additional blocking variables are required.

SAQs AND RESPONSES FOR PART TWO

SAQ 2.2a

Suppose that you have to examine the effects of three treatments (we shall call them A, B, C) with four replicates in twelve runs. Unfortunately the experimental procedure is lengthy and you can only manage three runs per day. Produce a randomized block design which will enable you to decide what effects the treatments have, whilst at the same time estimating the residual error and block (day-to-day) effect.

Response

The key to this design lies in having four replicates available to you. This means that you are able to examine each treatment once a day if you so wish. This will satisfy the need for balance between the blocks (days). However, you also have to decide upon an order for the treatments during the day, and this has to be produced in a random way to avoid the possibility of inflating the differences between the treatments. I chose to use a random number generator. First of all I randomized the day numbers (the first random numbers between 1 to 4) and then generated the order of the treatments per day (A = 1, B = 2, C = 3). The resulting design was:

Treatment

Day 1	A	C	B
Day 2	A	B	C
Day 3	C	B	A
Day 4	B	A	C

Please note that each treatment is tested once in each block.

****

SAQ 2.2b

> In an experiment examining the effect of six batches of material and four treatments in 48 experiments, how many degrees of freedom can be allocated to each of the sum of squares values?

Response

The *Total d.f.* is always given by the number of experiments but since we always want to know if there are significant differences between the treatment averages at a particular probability level then we have one less degree of freedom available and the Total *d.f.* is usually quoted as $N - 1$, which in this case is $48 - 1 = 47$. The *Block d.f.* are the number of blocks minus one $(b - 1) = 6 - 1 = 5$ and Treatment *d.f.* are the number of treatments minus one $(t - 1) = 4 - 1 = 3$. Finally the *Residual error d.f.* can be worked by subtraction from the *Total d.f.* [*Total d.f.* − $(t - 1)$ − $(b - 1) = 47 - 3 - 5 = 39$].

****

SAQ 2.2c	In an experiment to compare the percentage efficiency of four different chelating resins in extracting Cu^{2+} ions from aqueous solution, the experimenter decided to use a randomized block design since it was only possible to carry out four runs with each resin. On each of three successive days a fresh solution of the Cu^{2+} ions was prepared and the extraction was performed with each of the chelating resins, taken in a random order.

A randomized block design was set up and the following results were obtained:

	Chelating resin			
Day	A	B	C	D
1	97	93	96	92
2	90	92	95	90
3	96	91	93	91
4	95	93	94	90

Decide by an appropriate analysis whether the differences between the resins and the days on which the analyses were conducted had significant effects at the $P = 0.05$ probability level.

Response

I hope that you quickly realised that the analysis requires you to set up a two-way table. I subtracted the average (93%) from each of the results before I did this so as to make the subsequent calculations easier [it makes the correction for the mean term (CM) equal zero]. You might have subtracted a different value or indeed not have subtracted anything. In any case your two-way table should be fairly similar to the one given below:

Block (Day)	Resin (Treatment)				Block total	$\left(\dfrac{\text{Block}}{\text{total}}\right)^2$	$\Sigma\left(\dfrac{\text{Block}}{\text{total}}\right)^2$
	A	B	C	D			
1	4	0	3	−1	6	36	62
2	−3	−1	2	−3	−5	25	
3	3	−2	0	−2	−1	1	
4	2	0	1	−3	0	0	
					Total (T) = 0		
Treatment total	6	−3	6	−9	$\Sigma y_{ijk}^2 = 80$		
$\left(\dfrac{\text{Treatment}}{\text{total}}\right)^2$	36	9	36	81			
$\Sigma\left(\dfrac{\text{Treatment}}{\text{total}}\right)^2$	162						

Therefore,   *TreatmentSS* = 162/4 = 40.5
and   *BlockSS* = 62/4 = 15.5

Using the additivity of the sums of squares terms to your advantage you should then have been able to work out the residual error sum of squares by difference.

$$\text{\emph{ResidSS}} = 80 - 40.5 - 15.5 = 24$$

You needed the numbers of degrees of freedom for each sum of squares term before you could go any further. The *Total d.f.* are the number of experiments minus one and therefore = 16 − 1 = 15. The *Treatment d.f.* = $t - 1 = 3$, the *Block d.f.* = $b - 1 = 3$, and the *Residual error d.f.* = 15 − 3 − 3 = 9.

You should then have been able to set up an analysis of variance table:

Source of variation	d.f.	SS	MS	Variance ratio
Treatment	3	40.5	13.33	5.00
Block	3	15.5	5.167	1.937
Residual	9	24	2.667	
Total	15	80		

If you worked these values out correctly (or even fairly approximately) all that remained was to look up the critical F values for the appropriate d.f. at $P = 0.05$. How many d.f. were there? I hope you knew automatically that the between-treatment to residual error variance ratio had three and nine d.f. Upon examining the F table with these d.f. you should have found a value of 3.86. This is less than the calculated value (5.00). You should therefore have stated with at least 95% confidence that there are differences between the amounts of metal extracted by the different resins. However, the calculated value for the blocks did not exceed this value and at a $P = 0.05$ probability level you could not state that the day on which any of the analyses were conducted had an effect upon the amount of Cu^{2+} extracted from the water.

If you had difficulty in performing the above calculations you might find it advantageous to re-read about *ANOVA* in Part 1 and to read the introductory sections of Part 2 again. They should follow on quite logically, and it is probably a good idea to clear up any problems you have with these analytical techniques before moving on to more complicated aspects of design.

****

SAQ 2.3

Complete the following paragraph by inserting the most appropriate word or phrase, chosen from the list given below, into the blank spaces.

By an experimental design we mean a planned series of operations called .... in which certain conditions are held as constant as possible throughout and others are varied under the close control of the experimenter. The conditions which are particular to one experiment comprise the .... When more than one variable is changed between experiments the individual conditions are called .... The particular values at which an experiment are run are called the factor .... and where identical treatments are possible .... is used. If the experimenter is unable to keep certain conditions for one factor constant over the whole series of experiments .... is used to make comparisons more reasonable. A simple design for such a situation is the .... design. Sometimes, however, the blocks are of insufficient size to accommodate all the treatments and the experimenter may have to resort to .... designs. When two blocking variables are unavoidable a .... design may be used and when three blocking variables are necessary a .... design can be used. .... is a technique which must be applied to all these designs as a means of avoiding invalid conclusions. However, these designs suffer from the disadvantage that many of the available degrees of freedom are used up in estimating the block sum of squares. For example a 4 × 4 graeco-latin square design has only .... d.f. available for the residual sum of squares.

$\longrightarrow$

SAQ 2.3
(cont.)

BLOCKING	FIVE	GRAECO-LATIN SQUARE
SIX	BALANCED COMPLETE BLOCK	TREATMENT
RANDOMIZED BLOCK	LEVELS	EXPERIMENTS
LATIN SQUARE	BALANCED COMPLETE BLOCK	
FACTORS SEVEN	REPLICATION	RANDOMIZATION

Response

By an experimental design we mean a planned series of operations called EXPERIMENTS in which certain conditions are held as constant as possible throughout and others are varied under the close control of the experimenter. The conditions which are particular to one experiment comprise the TREATMENT. When more than one variable is changed between experiments the individual conditions are called FACTORS. The particular values at which an experiment are run are called the factor LEVELS and where identical treatments are possible REPLICATION is used. If the experimenter is unable to keep certain conditions for one factor constant over the whole series of experiments BLOCKING is used to make comparisons more reasonable. A simple design for such a situation is the RANDOMIZED BLOCK design. Sometimes, however, the blocks are of insufficient size to accommodate all the treatments and the experimenter may have to resort to BALANCED INCOMPLETE BLOCK designs. When two blocking variables are unavoidable a LATIN SQUARE design may be used and when three blocking variables are necessary a GRAECO-LATIN SQUARE design can be used. RANDOMIZATION is a technique which must be applied to all these designs as a means of avoiding invalid conclusions. However, these designs suffer from the disadvantage that many of the available degrees of freedom are used up in estimating the block sum of squares. For example a 4 × 4 graeco-latin square design has only SIX *d.f.* available for the residual sum of squares.

This SAQ is intended to act as a summary of Part 2. If you have followed it reasonably well the SAQ should have presented you with few problems.

****

3. Factorial Designs and Analysis

In Part 2, single-factor designs with one or more blocking factors were introduced as a means of running experiments in which insufficient time or material is available to complete the experiment under reasonably constant conditions. The resulting analysis of variance allows you to separate the effects of the factor from the block effect(s). However, these designs are fairly limited in scope and we need to look at designs which are applicable when the effects of more than one factor are of equal interest. This will be the topic of discussion in this part.

3.1. THE FACTORIAL DESIGN (DESIGN WITH MORE THAN ONE FACTOR)

In contrast to single-factor designs which are sub-divided into homogeneous blocks, two factors may be of equal interest to you and you might wish to have information on both their effects. Let us take the example in which the analysis of soils for total nitrogen was examined. We decided in Part 2 that several variables might influence the amount determined and included, amongst others, the type of catalyst used and the digestion temperature. Suppose you wanted to investigate the effects of both these factors in one experiment and they are both of equal interest. You could set up the experiment so that all the possible combinations of the factors are used. A simple design for these two factors would have both factors at two levels (two catalysts, two digestion temperatures). In this case the catalyst is a qualitative factor because it can only be changed in a qualitative manner (in which either A or B is used). On the other hand, temperature is a quantitative factor since its value can be changed readily.

The results (% N_2) are

	Digestion temperature (°C)	
	230	260
Catalyst A	3.5	3.0
B	2.5	2.0

Examination of the information available indicates that the relationship between the % N_2 and digestion temperature can be approximated by a straight line for both catalysts (Fig. 3.1a). When two levels of a quantitative factor such as digestion temperature are used, the only functional form which you can fit is a straight line, and you must assume that, over the range of the factor studied, the relation between the expected responses and the levels of the factor is approximately linear.

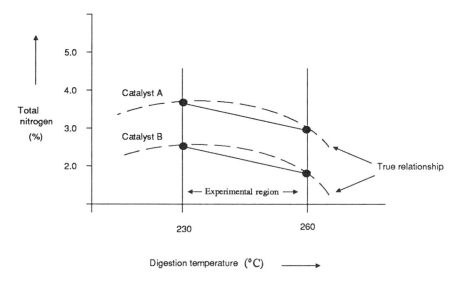

Fig. 3.1a. *Additive effect of digestion temperature and catalyst type*

The effect of changing the catalyst is the same at both digestion temperatures and is estimated as the average difference between the two lines in Fig. 3.1a;

$$\text{Catalyst effect} = \frac{[(2.5 - 3.5) + (2.0 - 3.0)]}{2} = -1.0\%$$

This is termed the *main effect* of the catalyst. Similarly the main effect of increasing the digestion temperature is the same when using either catalyst and may again be determined as the average difference:

$$\text{Temperature effect} = \frac{[(3.0 - 3.5) + (2.0 - 2.5)]}{2} = -0.5\%$$

The straight lines indicated in Fig. 3.1a may be perfectly adequate for your purpose (and you cannot guess that the underlying true response function is in fact slightly curved).

3.1.1. Interactions

If the results obtained (% N_2) had been as follows:

	Digestion temperature (°C)	
	230	260
Catalyst A	4.0	3.0
B	2.0	4.0

then the main effects could have been calculated as previously, giving + 0.5% for digestion temperature and −0.5% for catalyst. However, it is obvious that the effect of increasing the digestion temperature is different for the two catalysts, giving a decrease of 1.0% with catalyst A and an increase of 2.0% with catalyst B. There is said to be an *interaction* between these factors. Quite clearly a discussion of the main effects of the factors upon the experimental system could be very misleading unless some measure of the interaction between these factors is taken into account.

The interaction between these two factors can be estimated as half the difference between the effects of digestion temperature for the two catalysts. Thus for catalyst A the effect is 3.0 − 4.0 = −1.0% and for catalyst B is 4.0 − 2.0 = +2.0%. Therefore,

$$\text{Interaction effect} = \frac{[+2.0 - (-1.0)]}{2} = 1.50\%$$

This effect is fairly large in comparison to the main effects calculated above and should therefore be taken into account in any discussion of the effects. When the results are plotted (Fig. 3.1b), the extent of this interaction becomes clear, especially if you compare this plot with the parallel lines of Fig. 3.1a when no interaction was present. Therefore, when there is any possibility of an interaction between the factors, an experimental design is required which will estimate both the main effects and the interactions. Factorial designs are a method of doing this.

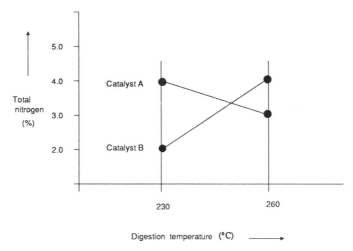

Fig. 3.1b. *Interactive effect of digestion temperature and catalyst type*

Re-examining Fig. 3.1a, you should see that it might be possible to estimate the extent of curvature (a quadratic effect) in the response by carrying out additional experimental runs at an intermediate digestion temperature. This curvature could then be estimated by the extent of deviation of the response at the intermediate level from a straight line connecting the outer two responses. A good design for this particular situation might be a 2 × 3 (mixed) factorial design in which the additional digestion temperature is set so that the three points are evenly spaced. This design allows you to estimate both the straight line (main effect) slope and the degree of curvature in the response.

In the above situation you could not easily have guessed the type of underlying response function. It might have been a straight line or indeed a more complex curved response. This, in a nutshell, is the basic problem of experimental design. How do you decide what pattern of design points will best reveal relevant aspects of the situation of interest?

If results are obtained at an intermediate level of digestion temperature, the extent of curvature in the response as a function of temperature might be clear for both catalysts. However, the spacing of the digestion temperatures might be too wide and tests at additional points may have to be run.

∏ Several soil samples were analysed for nickel in two laboratories. Suppose that one of the samples was more granular in nature than the others and had to be ground finely prior to the analytical determination. The results (μg g^{-1} of nickel) for the study are given below;

		Laboratory	
		A	B
Sample	1	10	12
	2	35	37
	3	5	7
(Granular)	4	21	?

What would you expect as the result for sample 4 in laboratory B with

(a) efficient grinding equipment?

(b) faulty grinding equipment which led to an incomplete extraction?

(a) If the grinding equipment was working correctly then the observation that all the results in laboratory B are 2 μg g^{-1} higher for each of the samples studied would also hold for sample 4. A reasonable answer would therefore be 23 μg g^{-1}.

(b) If the grinding equipment of laboratory B was faulty the extraction of nickel would have been incomplete and a low result recorded (e.g. 20 μg g^{-1}). This is, therefore, a laboratory–sample interaction since this effect only occurred for one of the samples in one of the laboratories.

∏   It is proposed to investigate the effects of three operational
       temperatures (100, 120 and 140 °C), two reactant concentrations
       (5 and 20%), and two catalysts (A and B) on the yield produced
       by a reactor vessel. What is the minimum number of experimental
       runs in which all the possible combinations of the factors could be
       investigated?

If your answer is 12 award yourself a gold star and move on because
you seem to understand how to work these out. If on the other hand your
answer does not concord you should read the following paragraph and it
may become clear. Otherwise go back to the previous section.

To investigate the effects of the three different temperatures at a constant
reactant concentration and catalyst type, a minimum of three experimental
runs are necessary. To investigate the effect of changing the reactant
concentration from 5% to 20% at a constant 100 °C, two runs are needed.
Similarly two runs are needed to investigate the effect of changing the
concentration from 5% to 20% at 120 °C and two runs for 140 °C. Thus
far a total of six runs is required without even considering the effect of the
catalyst type. However, the logic can easily be extended to investigate the
effect of changing the catalyst from A to B. One additional run is necessary
for each previous set of conditions, giving a total of 12 runs. This constitutes
a 2 × 2 × 3 factorial design.

∏   If the effects of four variables are to be studied, one at three levels
       and the others at two levels, how is the design designated? How
       many different runs are required?

This is a design of mixed factorial type and may be written either in a long
form according to each variable as:

$$3 \times 2 \times 2 \times 2$$

or as 3×2^3 since one variable is to be examined at three levels and the
others at two levels. The number of runs required can be calculated as a
product of the above = $3 \times 2 \times 2 \times 2 = 24$.

SAQ 3.1a   Match the words on the left-hand side to the most appropriate phrase on the right-hand side.

(a) factor   1) an observed numerical result

(b) levels   2) when the effects of two or more factors are not additive

(c) response   3) a combination of factor levels used in an experiment

(d) interaction   4) different values for a factor

(e) treatment   5) a variable believed to affect the outcome of an experiment

The responses to SAQs in Part 3 begin on page 133.

3.1.2. Advantages and Limitations to Factorial Designs

The advantages of factorial experiments can easily be shown for the earlier example in which only two digestion temperatures, T_0 and T_1, were considered for the two catalysts C_A and C_B. The minimum number of experimental runs necessary to give information on both factors is three, one at $T_0 C_A$, a second at $T_1 C_A$ involving a change in digestion temperature and a third at $T_0 C_B$ involving a change in catalyst only. These trials give the responses (y_1), (y_2) and (y_3) in the following table:

	Digestion temperature	
Catalyst	T_0	T_1
C_A	(y_1)	(y_2)
C_B	(y_3)	

The effect of changing the digestion temperature is given by $(y_2) - (y_1)$, and the effect of changing the catalyst is given by $(y_3) - (y_1)$. Thus, there is one comparison available for each main effect. Often, however, it is useful to have some confirmation of such effects, and duplication of each of the above treatments is one way of achieving this. The total number of runs is then 6, with the effects being calculated from the treatment averages. This is known as the one-variable-at-a-time approach since each factor is investigated separately.

Suppose now that the above table is completed by carrying out a trial with the treatment $T_1 C_B$, denoted by response (y_4). There are now two measures of the effect of changing the digestion temperatures, one with catalyst A as given by $(y_2) - (y_1)$ and the another with catalyst B as given by $(y_4) - (y_3)$. In the case that there is no interaction between the two factors, the estimates will differ only because of experimental error and may be averaged to yield the main effect of changing the digestion temperature. Similarly there are two measures of the effect of changing the catalyst, one at each of the digestion temperatures as calculated from $(y_3) - (y_1)$ and $(y_4) - (y_2)$ respectively. These estimates are based on only four experimental runs as compared to the six required for the one-variable-at-a-time approach, a saving of two experimental runs, and yet are still based on duplicate comparisons.

An additional advantage is that variables often interact. The one-variable-at-a-time approach cannot estimate the magnitude of interactions and might often miss important conclusions about the dependence of the effect of one factor on the level of another factor.

Where a main effect is apparently large and interactions are relatively small, the main effect has been examined over differing levels of other factors and could indicate a substantially true or real effect. Factorial designs can also be very useful in the process of optimization since estimates of the main effects and interactions can be used to predict the likely position of an optimum combination of the factors. This will be explored further in Part 5.

However, a word of caution is appropriate at this point. Factorial experiments are not a panacea for experimental design. In some cases it is simply not possible to use them because some of the treatment combinations are not possible for one reason or another. Absence of treatment combinations from a factorial design will lead to an unbalanced design in which some effects cannot be estimated separately. They also tend to require a large number of runs even when the number of factors is fairly small. For example, an unreplicated 3^3 factorial design requires 27 runs. This point will be explored further in Part 4.

3.1.3. Sequential Approach to Experimental Investigation

When exploring a functional relationship between a response and a number of factors, it might appear reasonable to adopt a comprehensive approach and try to investigate the entire range of every factor. To carry out all the possible combinations of the factors would then involve a great experimental effort. However, you can often carry out runs in successive groups, using the information gained in the preliminary stages to decide which treatments will be used in the later experimental runs. For example, some of the factors might well prove to be unimportant and can be excluded from the later runs, so reducing the size of the experiment. Also the experiment may be moved to a more promising region (optimization) which can be explored very thoroughly. In addition the information gained may indicate that the variables should be transformed, e.g. converted to logarithms, and the levels adjusted accordingly.

As a rough guide you should not use more than one quarter of the budgeted experimental effort in the first design. Of course there are exceptions to this rule, especially influenced by the size of the batches of any material used, time considerations etc. Nevertheless it is generally unwise to plan too comprehensive an initial design. When a preliminary investigation has been completed you will usually know more than before and may be able to plan a better second part. Information from this may be used to plan a better third part and so on.

SAQ 3.1b

> In an inter-laboratory study several samples are to be analysed by different methods at different temperatures. How many different effects should you take into account?

SAQ 3.1c

For an unreplicated 3^4 design

1. How many factors (variables) are to be studied?

2. How many levels are these factors to be studied at?

3. What is the total number of runs required?

3.2. FACTORIAL DESIGN EXPERIMENTS AT TWO LEVELS

3.2.1. Introduction

To carry out a general factorial design, you select a certain number of levels for each factor (variable) and then run an experiment with all the possible combinations of the factors. You should have learned in the previous section that the total number of runs can be obtained by multiplication of the levels for each of the factors. For example, a $2 \times 3 \times 4$ mixed factorial requires 24 runs.

One class of factorial design which has found some favour with some experimenters is the three-level factorial (3^f) design, where f is the number of factors. Such a design can provide information on the effects and interactions of the factors in which main or linear effects and quadratic or curvature effects can be separated. It is often quite difficult to interpret this information. Also 3^f designs require considerable numbers of experimental runs. In this section the simplest and the most popular class of factorial design in which all the factors have only two levels is introduced, and is denoted as the 2^f class. One reason for the popularity of these designs with experimenters is that they require relatively few runs per factor studied. The results of the designs are also relatively easy to interpret and, although they do not investigate a wide region of the factor space, can indicate major trends. They also form the basis of fractional factorial designs (Part 4) and can be augmented to form composite designs for the investigation of response surfaces (Part 5).

3.2.2. Example

To demonstrate specific points about the design and analysis of two-level factorial experiments, an example investigation is now introduced and will be referred to in the following section.

In the separation of phenols by reverse phase high performance liquid chromatography (HPLC), several chemical factors are known to influence the separation. The aim of this study is to examine which factors are important and to optimize the separation of 11 priority pollutant phenols on an isocratic HPLC system, where isocratic means that the solvent composition cannot be altered during the chromatographic run.

Several chemical factors are known to influence the separation and those to be studied are the proportion of methanol (M) in the mobile phase (methanol:water) and the concentrations of citric acid (C) and acetic acid (A) in the mobile phase. Citric acid and acetic acid are added to the mobile phase because they can reduce the degree of peak tailing, a severe phenomenon in the HPLC of phenols. Therefore, all the variables are quantitative. The response to be monitored is the chromatographic response function (CRF), a summation term of the individual resolutions between pairs of peaks. Thus, if the peaks are separated at the base-line and the degree of peak tailing is small the CRF will have a high value. The design used is one of two levels, a 2^3 design requiring a total of 8 runs.

In any study with the factors at two levels it is possible to assign − and + signs to indicate which level of the factor is applied in any particular treatment. For a quantitative factor the − and + signs will conveniently represent the factor at its lower and higher level respectively. For a qualitative factor, the two versions or "levels" can also be coded by − and + signs, although it is unimportant which is the − sign as long as the labelling is consistent. The levels used for this experiment are:

	−	+
Acetic acid concentration (mol dm^{-3})	0.004	0.01
Methanol (%)	70	80
Citric acid concentration (g dm^{-3})	2	6

3.2.3. Notation

A display of the levels to be included in a design is called a design matrix, and three representations of the same design matrix for the HPLC example are given below:

Run	A	M	C		A	M	C
1	−	−	−	*(1)*	0	0	0
2	+	−	−	*a*	1	0	0
3	−	+	−	*m*	0	1	0
4	+	+	−	*am*	1	1	0
5	−	−	+	*c*	0	0	1
6	+	−	+	*ac*	1	0	1
7	−	+	+	*mc*	0	1	1
8	+	+	+	*amc*	1	1	1

The matrix on the left-hand side consists of − and + signs denoting the
levels of each factor in each of the runs. Thus run 4 is represented by +
+ − for factors *A*, *M* and C respectively, indicating that acetic acid and
methanol are used at their high levels and citric acid at its low level. It is
also common practice to denote a factor by a capital letter, and the higher
level of any factor in a particular treatment by the corresponding lower
case letter. For example *am* is the treatment in which factors *A* and *M*
are at their high levels and C at its low level. The treatment where the
factors are all at their lower levels (run 1) is represented by *(1)*. If the
factor is qualitative in nature then the absence of the lower case letter in
a particular treatment is often used to represent the normal condition (if
there is one). The lower case letter is then used to represent the changed
condition.

The third notation used on the right-hand side, consisting of 0's and 1's for
each treatment, is synonymous with the − + notation used on the left-hand
side with 0 being equivalent to − and 1 being equivalent to + . Please note
that the term *Run* refers to the standard order of specifying the treatments
and not to the order in which the runs are carried out (which should of
course be random).

The order of the treatments was actually randomized using one of the techniques previously described and the experimental runs were carried out. The *CRF* (y) values obtained are given below along with a complete design matrix with coded variables.

Run number	Acetic acid A	Methanol M	Citric acid C	CRF value (y)
1	−	−	−	10.0
2	+	−	−	9.5
3	−	+	−	11.0
4	+	+	−	10.7
5	−	−	+	9.3
6	+	−	+	8.8
7	−	+	+	11.9
8	+	+	+	11.7

Π   What was the *CRF* value obtained at an acetic acid concentration of 0.004 mol dm^{-3}, a methanol percentage of 80% and a citric acid concentration of 6 g dm^{-3}?

The question asked for the *CRF* value at an acetic acid concentration (A) of 0.004 mol dm^{-3}. This limits the choice to one of runs 1, 3, 5 or 7 which are all the runs carried out at 0.004 mol dm^{-3} as indicated by the − signs under A. However, the required percentage of methanol was 80% (+) and the only runs in which A was − and M was + are runs 3 and 7. Run 3 had a citric acid concentration of 2 g dm^{-3} (−). Therefore, the correct *CRF* value is that of run 7 which is 11.9.

SAQ 3.2

Given below is a 2^4 factorial design to examine the effects of four factors, current (C), ratio of time of plating ON to OFF (R), strength of the acid solution (A), and frequency of the plating signal (F), upon an electroplating procedure. Give two alternatives to the notation used.

Factor			
C	R	A	F
−	−	−	−
+	−	−	−
−	+	−	−
+	+	−	−
−	−	+	−
+	−	+	−
−	+	+	−
+	+	+	−
−	−	−	+
+	−	−	+
−	+	−	+
+	+	−	+
−	−	+	+
+	−	+	+
−	+	+	+
+	+	+	+

3.2.4. Graphical Representations

Two-level factorial designs can be represented graphically. This is generally helpful in the interpretation of the responses. Where the effects of two

factors are to be studied a square may be used, as given below in Fig. 3.2a
for the investigation into the effects of catalyst and digestion temperature
upon % N_2. The numbers on the points of the square are the responses
obtained.

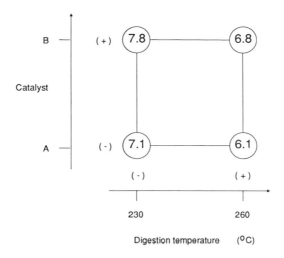

Fig. 3.2a.   *2^2 factorial design, Kjeldahl analysis (response = % nitrogen)*

∏   The % N_2 obtained when both factors C and T are at low levels is
       7.1%; what are the levels of C and T that give 6.8% N_2?

You should be able to see from the diagram that 6.8% N_2 was obtained
at the + levels for both factors since it is diagonally opposite the 7.1% N_2
obtained when both factors are at their low levels.

Where there are three or more factors it is possible to represent the factors
as faces on one or more cubes with the responses at the points. This is
demonstrated in Fig. 3.2b for the 2^3 HPLC experiment.

Where there are four factors, one cube may represent the fourth factor
at its lower level (or where a qualitative factor has one condition), and
a second cube may represent the fourth factor at its higher level (or a
qualitative factor at another condition). The number of cubes may be

doubled for every additional factor, but this approach is less useful at higher numbers of factors because it becomes increasingly difficult to envisage the effects.

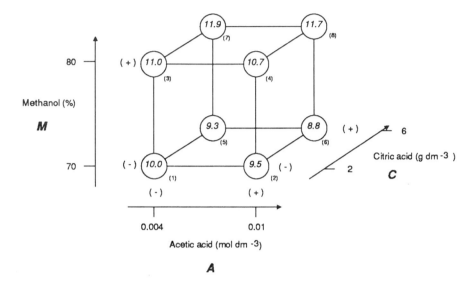

Fig. 3.2b. 2^3 *factorial design, HPLC example (response = CRF)*

3.3.   ESTIMATION OF EFFECTS

It may be argued that in this age of computers there is very little need to understand the mechanism by which effects are calculated, but without at least a cursory understanding of the process it will be hard for you to appreciate exactly what a factorial design can do and indeed what its advantages and limitations may be. The example used is that of the 2^3 HPLC experiment.

3.3.1.   Main Effects

When attempting to calculate effects it is sometimes useful to view the full treatment combinations in both their original and coded values, along with the responses obtained for each of the runs.

Run number	Acetic acid (mol dm^{-3}) A	Methanol (%) M	Citric acid (g dm^{-3}) C	CRF value y
	Original variables			
1	0.004	70	2	10.0
2	0.01	70	2	9.5
3	0.004	80	2	11.0
4	0.01	80	2	10.7
5	0.004	70	6	9.3
6	0.01	70	6	8.8
7	0.004	80	6	11.9
8	0.01	80	6	11.7
	Coded variables			
1	−	−	−	10.0
2	+	−	−	9.5
3	−	+	−	11.0
4	+	+	−	10.7
5	−	−	+	9.3
6	+	−	+	8.8
7	−	+	+	11.9
8	+	+	+	11.7

One estimate of the effect of changing the concentration of the acetic acid upon the *CRF* when both the other factors are constant (*M* and *C* are −) is given by the difference in responses between experimental runs 1 and 2. Similarly, further estimates of the acetic acid effect are obtained from the differences between the responses of runs 3 and 4 (*M* is +, *C* is −) and so on for runs 5 and 6, and 7 and 8. Thus four estimates of the effect of changing the acetic acid concentration are available and an average of these may be calculated.

$$\frac{\text{acetic acid}}{\text{effect}} = \frac{(y_2 - y_1) + (y_4 - y_3) + (y_6 - y_5) + (y_8 - y_7)}{4}$$

$$= \frac{(9.5 - 10.0) + (10.7 - 11.0) + (8.8 - 9.3) + (11.7 - 11.9)}{4}$$

$$= \frac{(-0.5) + (-0.3) + (-0.5) + (-0.2)}{4} = -0.375$$

Therefore, increasing the acetic acid concentration from 0.004 to 0.01 mol dm^{-3} has had an negative effect upon the chromatographic response function (CRF) of approximately -0.4 unit. Note that I have used all eight experimental runs to calculate this effect.

The acetic acid effect could also have been calculated by subtracting the average of the responses with A at its lower level (0.004 mol dm^{-3}) from the average of the responses with A at its higher level (0.01 mol dm^{-3}).

$$\frac{\text{acetic acid}}{\text{effect}} = \frac{y_2 + y_4 + y_6 + y_8}{4} - \frac{y_1 + y_3 + y_5 + y_7}{4}$$

$$= \frac{9.5 + 10.7 + 8.8 + 11.7}{4} - \frac{10.0 + 11.0 + 9.3 + 11.9}{4}$$

$$= 10.175 - 10.55 = -0.375$$

Π   From the data given above for the HPLC example, calculate the effects of changing the percentage methanol (M) from 70% to 80% and changing the citric acid concentration from 2 to 6 g dm^{-3}.

The answer you should have obtained for the effect of changing M is 1.925, which is obtained by subtracting the average of the responses with M at its lower level $(-)$ from the average of the responses with M +;

$$\text{methanol effect} = \frac{y_3 + y_4 + y_7 + y_8}{4} - \frac{y_1 + y_2 + y_5 + y_6}{4}$$

$$= \frac{11.0 + 10.7 + 11.9 + 11.7}{4} - \frac{10.0 + 9.5 + 9.3 + 8.8}{4}$$

$$= 11.325 - 9.4 = 1.925$$

Thus the highest responses were obtained with the methanol set to 80%.

Similarly I calculated the citric acid effect by subtracting the average response with C at its lower level ($-$) in runs 1, 2, 3 and 4 from the average at its higher level ($+$) in runs 5, 6, 7 and 8.

$$\begin{array}{c}\text{citric acid} \\ \text{effect}\end{array} = \frac{y_5 + y_6 + y_7 + y_8}{4} - \frac{y_1 + y_2 + y_3 + y_4}{4}$$

$$= \frac{9.3 + 8.8 + 11.9 + 11.7}{4} - \frac{10.0 + 9.5 + 11.0 + 10.7}{4}$$

$$= 10.425 - 10.3 = 0.125$$

This effect is relatively small in comparison to those of A and M. There has clearly been a great saving in the number of runs used over the number that would have been required if the factors had been varied one at a time since all the experimental runs have been used to calculate each of the main effects, allowing four comparisons to be made for each. Four replicates of each of the levels would have been carried out to provide estimates with the same degree of precision.

Cube diagrams are also very useful in obtaining the estimates of the main effects since each face represents a factor at one of its levels. To calculate any main effect subtract the average of the responses on the face with the factor at its lower level ($-$) from the average on the opposite face ($+$). In the 2^3 HPLC example there are three factors each at two levels corresponding to the six faces of a cube. Thus to calculate the acetic acid effect average the responses on the left face and subtract the result from the average of the responses on the right face as shown in Fig. 3.3a.

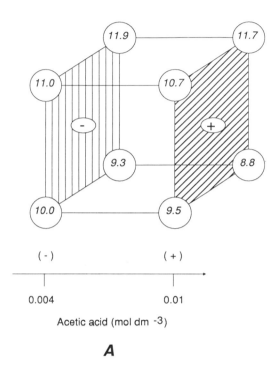

A

Fig. 3.3a. *2^3 factorial design, acetic acid effect, HPLC example (response = CRF)*

Similarly, to calculate the methanol effect subtract the average response on the bottom face from that on the top face and to calculate the citric acid effect use the front and back faces.

3.3.2. Interactive Effects

In this example the average effect of methanol is 1.93. From the cube shown in Fig. 3.3b with the differences between the lower and higher levels added, it is clear that the methanol effect is much greater with the C set to 6 g dm^{-3} than it is with C set to 2 g dm^{-3}, i.e. M and C do not behave additively and can be said to interact.

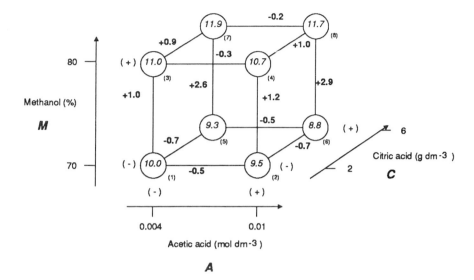

Fig. 3.3b. *2³ factorial design, HPLC example showing differences between levels of factors (response = CRF)*

A measure of this interaction is supplied by half the difference obtained when the average methanol effect with citric acid − is subtracted from the average methanol effect with citric acid +. By convention this is called the methanol by citric acid interaction (*MC*):

citric acid	average methanol effect	
(+) 6 g dm⁻³	(2.9 + 2.6)/2 =	2.75
(−) 2 g dm⁻³	(1.2 + 1.0)/2 =	1.1
	difference	1.65

$MC = 1.65/2 = 0.825$

It may equally be thought of as one-half the difference in the average citric acid effects at the two levels of methanol and can be calculated accordingly:

methanol	average citric acid effect	
(+) 80%	$(1.0 + 0.9)/2$	$= \ \ 0.95$
(−) 70%	$(-0.7 - 0.7)/2$	$= -\underline{0.7}$
	difference	1.65

$$MC = 1.65/2 = 0.825$$

∏ Again using the data from the HPLC experiment, calculate the effects of the two-factor interaction between acetic acid and citric acid (AC), and the interaction between acetic acid and methanol (AM).

In order to calculate the AC interaction you should have considered that there are two estimates for the acetic acid effect, one at each level of citric acid. Half the difference between these estimates is the AC interaction.

Thus the effect of acetic acid with citric acid at 6 g dm^{-3}

$$= \frac{(y_8 - y_7) + (y_6 - y_5)}{2} = \frac{(11.7 - 11.9) + (8.8 - 9.3)}{2}$$

$$= -0.35$$

The effect of acetic acid with citric acid at 2 g dm^{-3}

$$= \frac{(y_4 - y_3) + (y_2 - y_1)}{2} = \frac{(10.7 - 11.0) + (9.5 - 10.0)}{2}$$

$$= -0.4$$

Therefore,

$$AC \text{ effect } = \frac{(-0.35) - (-0.4)}{2} = 0.05/2 = 0.25$$

You could have obtained a numerically equivalent answer by considering the consistency of the average effects of C at the two levels of A.

Similarly you could have calculated the AM interaction by considering the average methanol effects at the two acetic acid levels:

acetic acid   average methanol effect

$+ (0.01 \text{ mol dm}^{-3}) \quad = \quad \dfrac{(y_4 - y_2) + (y_8 - y_6)}{2}$

$\qquad\qquad\qquad\quad = \quad \dfrac{(10.7 - 9.5) + (11.7 - 8.8)}{2}$

average effect of M with $A + = 2.05$

$- (0.004 \text{ mol dm}^{-3}) \quad = \quad \dfrac{(y_3 - y_1) + (y_7 - y_5)}{2}$

$\qquad\qquad\qquad\quad = \quad \dfrac{(11.0 - 10.0) + (11.9 - 9.3)}{2}$

average effect of M with $A - = 1.8$

$AM \; effect = \dfrac{2.05 - 1.8}{2} = 0.125$

Alternatively you could have obtained the same answer if you had examined the consistency of the acetic acid effect at the two methanol levels.

When the main effects were calculated (Section 3.3.1) the treatments were considered to be in two halves with the factors at their lower levels in one half and the higher levels in the other. Having made this division, the responses obtained for these treatments were averaged and the average at the lower level subtracted from the average at the higher level. It is also possible to consider any interactive effect in the same way. In the 2^3 HPLC design each interactive effect may be calculated from a difference between two averages with half the eight in one average which can be considered as the lower level for that interaction. The remaining treatments will thus make up the other half which can be considered as the higher level for the interaction. Arithmetically it is possible to rearrange the above calculations of the two-factor interactions to show this. For example the AM effect can be calculated as below:

$$AM \text{ effect} = \frac{y_1 + y_4 + y_5 + y_8}{4} - \frac{y_2 + y_3 + y_6 + y_7}{4}$$

$$= \frac{10.0 + 10.7 + 9.3 + 11.7}{4} - \frac{9.5 + 11.0 + 8.8 + 11.9}{4}$$

$$AM \text{ effect} = 10.425 - 10.3 = 0.125$$

Also, like the main effects, it is possible to view the two-factor interactions as contrasts between observations on a cube. However, instead of the cube faces which were used for the main effects, the contrasts are provided by diagonal planes within the cube (Fig. 3.3c). There is, for example, a diagonal plane comprising the *AM* effect at its lower level going from bottom right to top left in Fig. 3.3c (*i*). The responses on this plane may be averaged and subtracted from the average of the plane going from bottom left to top right. A brief examination of the runs should reveal that they are the same as used above.

∏   Using the cube diagrams given in Fig. 3.3c decide which runs are used as the higher level for the *AC* interaction.

I hope you can see that there are four runs for each of these averages. The diagonal plane used for the higher level is from the front left to back right and is composed of runs 1, 3, 6 and 8. Obviously the remaining four runs (2, 4, 5 and 7) make up the average for the lower level of *AC*.

Now consider the acetic acid by methanol (*AM*) interaction. Two measures of this are available, one for each level of citric acid;

AM interaction with *C* (+)

$$= \frac{(y_8 - y_7) - (y_6 - y_5)}{2} = \frac{(11.7 - 11.9) - (8.8 - 9.3)}{2} = 0.15$$

AM interaction with *C* (−)

$$\frac{(y_4 - y_3) - (y_2 - y_1)}{2} = \frac{(10.7 - 11.0) - (9.5 - 10.0)}{2} = 0.1$$

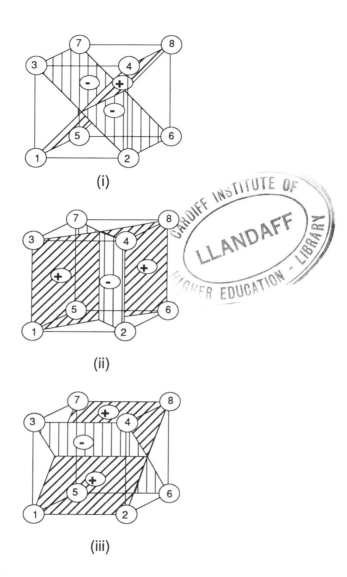

Fig. 3.3c. *2³ Factorial design, HPLC example. (i) AM interaction, (ii) AC interaction, (iii) MC interaction*

The difference between these two-factor interactions measures the consistency of the acetic acid by methanol interaction for the two levels of citric acid, and half this difference is defined as the three-factor interaction of acetic acid, methanol and citric acid;

$$AMC \text{ interaction } = \frac{(0.15) - (0.1)}{2} = 0.025$$

∏   Suggest at least one other way in which the three-factor interaction may be calculated.

There are six possible ways to calculate the *AMC* three-factor interaction. As shown above there are two ways of calculating each two-factor interaction and there are three of these in a 2^3 design, giving a total of six. For example it could be calculated from one-half the difference between the *AC* interaction at the two levels of methanol. The *AMC* three-factor interaction also has a certain geometric representation on the cube with the contrasts being provided by two tetrahedra. This is shown in Fig. 3.3d.

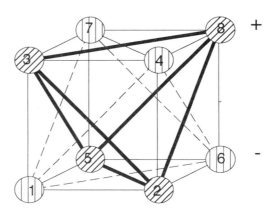

Fig. 3.3d. *2^3 Factorial design, HPLC example, AMC interaction*

SAQ 3.3a

Complete the following paragraph by inserting the most appropriate word or phrase, chosen from the list given below, into the blank spaces.

Two-level factorial designs estimate not only .... effects but also .... effects with maximum .... The main effects may be estimated as the difference between two ...., one with the .... at its + level, the other at its − level. Two-factor interactive effects may be estimated from half the .... between an average main effect at the .... level of another factor subtracted from the average effect at the ...... level. The .... of the effects may be judged from an estimate of the .... of an effect obtained by genuine replication when available.

HIGHER	AVERAGES
TWO-LEVEL	MAIN
INTERACTIVE	FACTOR
ACCURACY	RESULTS
VARIANCE	MAGNITUDE
PRECISION	SIGNIFICANCE
STANDARD ERROR	DIFFERENCE
LOWER	

3.3.3. Columns of Contrast Coefficients

It would be extremely tedious for you to work your way completely through a factorial design to calculate the interactive effects from first principles every time you came across one. There is a short cut provided by columns of contrast coefficients. These can be constructed from the original design matrix. This is shown below for the HPLC experiment along with the *CRF* values obtained:

Run	A	M	C	*CRF* (y)
1	−	−	−	10.0
2	+	−	−	9.5
3	−	+	−	11.0
4	+	+	−	10.7
5	−	−	+	9.3
6	+	−	+	8.8
7	−	+	+	11.9
8	+	+	+	11.7

The contrasts for the main effects are the signs used in the design matrix. These are applied to the responses and then divided by 4, since this is the number of comparisons made.

Thus, to calculate the acetic acid effect:

$$A = \frac{-y_1 + y_2 - y_3 + y_4 - y_5 + y_6 - y_7 + y_8}{4}$$

$$= \frac{-10.0 + 9.5 - 11.0 + 10.7 - 9.3 + 8.8 - 11.9 + 11.7}{4}$$

$$= -0.375$$

The other main effects could be calculated in a similar manner. Note that each effect has four − signs and four + signs associated with it.

Having given you the contrasts for the main effects you may now obtain contrasts for the interactive effects by directly multiplying the signs for the individual signs as shown below:

Run	A	M	AM
1	−	−	+
2	+	−	−
3	−	+	−
4	+	+	+
5	−	−	+
6	+	−	−
7	−	+	−
8	+	+	+

In run 1 the + sign for the *AM* two-factor interaction is obtained by multiplication of the − sign under A and the − sign under *M*. Similarly in run 2 the *AM* interaction can be regarded as at its low level and will have a − sign as its contrast since it is calculated from the multiplication of a − sign (*M*) and a + sign (*A*).

∏ The 2^3 factorial design matrix used in the HPLC investigation is given below. Which of columns 1, 2, 3 or 4 represents the interaction between variables *M* and *C*?

Run	A	M	C	1	2	3	4
1	−	−	−	+	+	+	−
2	+	−	−	−	+	+	−
3	−	+	−	+	+	−	+
4	+	+	−	−	−	−	+
5	−	−	+	−	+	−	+
6	+	−	+	+	−	−	+
7	−	+	+	−	−	+	−
8	+	+	+	+	+	+	−

If you thought that it could possibly be more than one of the choices then you are in need of some further revision on factorial designs and columns of contrast coefficients. The reason for this is that any interaction effect is a comparison of half the experiments (−) with the other half (+). These signs are the products of the individual signs composing the interactions and are thus unique.

If you chose column 1 as your answer then you were wrong because the *MC* interaction is composed of the individual signs multiplied together. Therefore, run 1 should have a + sign (product of a − multiplied by another −) as there is. However, in run 2 there should also be a + sign which there is not.

If, on the other hand, you chose column 2 you made a big mistake because you failed to notice that there were five + signs and three − signs. This could not be correct since every effect in a 2^3 complete factorial design, be it a main or interactive effect, is calculated as the difference between two sets of four experimental responses. It would also have been incorrect because of the + sign in run 3 where the *MC* interaction dictates that there should be a − sign.

If you chose column 3, well done! You obviously know how to multiply the signs to yield the interaction contrasts.

Column 4 could not be correct because for run 4 there is a − sign where the − sign under *M* multiplied by the − sign under *C* gives a + sign. A complete table of contrast coefficients for the 2^3 design is given below:

Mean	A	M	C	AM	AC	MC	AMC	CRF (y)
+	−	−	−	+	+	+	−	10.0
+	+	−	−	−	−	+	+	9.5
+	−	+	−	−	+	−	+	11.0
+	+	+	−	+	−	−	−	10.7
+	−	−	+	+	−	−	+	9.3
+	+	−	+	−	+	−	−	8.8
+	−	+	+	−	−	+	−	11.9
+	+	+	+	+	+	+	+	11.7
8	4	4	4	4	4	4	4	Divisor

Note that the signs for the *AMC* three-factor interaction can be obtained by multiplying the signs for any two-factor interaction by the individual sign for the third factor. For example, the + sign in run 1 for the *AM* interaction multiplied by the − sign for factor *C* yields a − sign for the *AMC* interaction and so on. To calculate the estimate for any interactive effects you simply apply the contrasts to the responses and divide by 4 as shown in Section 3.3.1 for the main effects. Thus,

$$AM = \frac{+ y_1 - y_2 - y_3 + y_4 + y_5 - y_6 - y_7 + y_8}{4}$$

$$= \frac{+10.0 - 9.5 - 11.0 + 10.7 + 9.3 - 8.8 - 11.9 + 11.7}{4}$$

$$= 0.125$$

SAQ 3.3b

Given below are the results of a 2^4 factorial experiment to examine the effects of four factors, current (C), ratio of time of plating ON to OFF (R), strength of the acid solution (A), and frequency of the plating signal (F), upon an electroplating procedure. Each of the factors was investigated at two levels, − and +. Use columns of contrast coefficients to calculate the main effects and interactions of the factors. The response was the hardness of the plate.

Factor				
C	R	A	F	Hardness
−	−	−	−	40
+	−	−	−	42
−	+	−	−	38
+	+	−	−	41
−	−	+	−	52
+	−	+	−	54
−	+	+	−	48
+	+	+	−	49
−	−	−	+	46
+	−	−	+	48
−	+	−	+	44
+	+	−	+	47
−	−	+	+	58
+	−	+	+	60
−	+	+	+	54
+	+	+	+	55

SAQ 3.3b

SAQ 3.3c

In an investigation into extraction of nitrate-nitrogen from air-dried soil, three quantitative variables were investigated at two levels. These were the amount of oxidized activated charcoal (A) added to the extracting solution to remove organic interferences, the strength of the $CaSO_4$ extracting solution (C), and the time that the soil was shaken with the solution (T). The aim of the investigation was to optimize the extraction procedure. The levels of the variables are given below:

		−	+
Activated charcoal (g)	A	0.5	1.0
$CaSO_4$ (%)	C	0.1	0.2
Time (minutes)	T	30	60

The concentrations of nitrate-nitrogen were determined by ultra-violet spectrophotometry and compared with concentrations determined by a standard technique. The results given below are the amounts recovered (expressed as the percent of the known nitrate concentration).

Run	A	C	T	Percent total (y)
1	−	−	−	68
2	+	−	−	65
3	−	+	−	75
4	+	+	−	72
5	−	−	+	90
6	+	−	+	89
7	−	+	+	99
8	+	+	+	95

Calculate and interpret the main effects of the factors and all the possible two- and three-factor interactions.

SAQ 3.3c

3.3.4. Yates' Method

A quick method of calculating the effects in factorial designs has been
provided by Yates, who developed an algorithm which is applicable to
both complete and fractional factorial designs (Part 4). To carry out Yates'
algorithm, you have to arrange the results in the standard order which you
have come across earlier. For a 2^3 factorial design the method is shown in
Fig. 3.3e. The responses for the corresponding experimental runs are placed
in column (y). From here on you consider these results in successive pairs.
You obtain the first four entries in the next column (1) by adding the pairs
of column (y) together. Thus $10.0 + 9.5 = 19.5$, $11.0 + 10.7 = 21.7$ and so
on. You then obtain the bottom four numbers of column (1) by subtracting
the top number of each pair from the bottom number. Thus $8.8 - 9.3 =
-0.5$ and $11.7 - 11.9 = -0.2$ and so on.

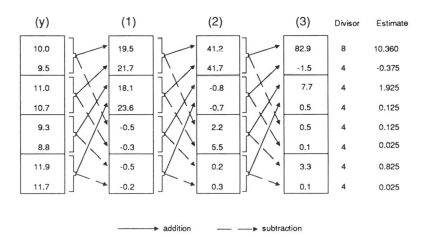

Fig. 3.3e. *Yates' algorithm, 2^3 HPLC example*

In the same way that column (1) is obtained from column (y), you obtain
column (2) from column (1). Finally column (3) is obtained from column
(2). You carry out the process of adding and subtracting the pairs of numbers
for as many times as there are factors. Thus, for the 2^3 design there are
three factors and therefore three columns produced in this way:

Run	A	M	C	(y)	(1)	(2)	(3)	Divisor	Estimate	Sum of squares
1	−	−	−	10.0	19.5	41.2	82.9	8	10.363	—
2	+	−	−	9.5	21.7	41.7	−1.5	4	−0.375	0.28125
3	−	+	−	11.0	18.1	−0.8	7.7	4	1.925	7.41125
4	+	+	−	10.7	23.6	−0.7	0.5	4	0.125	0.03125
5	−	−	+	9.3	−0.5	2.2	0.5	4	0.125	0.03125
6	+	−	+	8.8	−0.3	5.5	0.1	4	0.025	0.00125
7	−	+	+	11.9	−0.5	0.2	3.3	4	0.825	1.36125
8	+	+	+	11.7	−0.2	0.3	0.1	4	0.025	0.00125

The first number in the column (3) is the total of all the responses. This can be used as a check that the calculations have been performed correctly. On the right of the column (3) are the divisors for the estimated effects. The first divisor is therefore the number of experiments (8 in this case). All the other divisors are half this number. Therefore, a further column giving all the estimated effects is produced by dividing the number in the last column by its divisor. Now, I hear you say, which estimates are which? Well, you had to write the responses in the standard order from the design matrix and it is simply a matter of locating the + signs in this matrix. Thus the estimated effect of A is given in the second row because this is the only row in the design matrix which has a + sign under A alone.

∏   What is the row number which signifies the MC interaction?

The correct answer is row 7 which contains the + signs under M and C, whereas A is at its low level. Yates' method can also provide the sum of squares values if you want an analysis of variance. These are obtained by squaring the values in the final column (column 3 in this case) and dividing by the number of experiments. Thus, the sum of squares for $A = (-1.5)^2/8 = 0.28125$. The sum of squares for $M = 7.7^2/8 = 7.41125$ and so on.

You can easily enter these calculations on to a computerized spreadsheet (commercial templates are available). This may be quite useful to you when you have a large number of analyses to carry out and may be done with the same spreadsheet. These spreadsheets also have the advantages of providing you with a permanent record of your results and will be less susceptible to manual errors.

Once you have all the sum of squares values it is possible to construct an *ANOVA* table. At this stage you will need to allocate the degrees of freedom appropriately. Each effect has 1 d.f. summing to give a total of 7 ($N - 1$ = 7) and you can use them for the mean square estimates. However, in this experiment there was no replication of the treatments and all the *d.f.* have been allocated. What can you now use as an estimate of residual error variance?

Perhaps you can answer this question for your own laboratory situation. You may well be familiar with the process you are examining in the experiment and have a suitable estimate of residual error at hand. Perhaps you have documented repeated responses of identical experimental runs which differ only because of random variation. This may suffice. However, if you aren't so fortunate you may have to sacrifice your estimates of some of the higher-order interactions such as two- and three-factor interactions, pooling their sums of squares together and using this as an estimate of residual error variance. This may not be critical if the interactions are unimportant, but may be important if they are large. There are several reasons for this. Firstly, you will miss testing their significance and, secondly, they will tend to inflate the residual error sums of squares and could cause some effects to appear insignificant (statistically speaking) whereas in reality they are important. In this example I have pooled all the interaction sum of squares with their total of four d.f. giving a mean square value of 0.34875.

Source of variation	d.f.	SS		MS	Variance ratio
A	1	0.28125		0.28125	0.81
M	1	7.41125		7.41125	21.25*
C	1	0.03125		0.03125	0.09
AM	1 ⎫	0.03125 ⎫			
AC	1 ⎪ 4	0.00125 ⎬ 1.395	0.34875		
MC	1 ⎬	1.36125 ⎪			
AMC	1 ⎭	0.00125 ⎭			
Total	7				

* Exceeds the tabulated *F* value at the *P* = 0.05 probability level for one and four *d.f.* (7.71).

The complete *ANOVA* table reveals that only the *M* main effect is significant at the $P = 0.05$ probability level.

SAQ 3.3d Continuing with the plating example introduced in the previous SAQ, use Yates' method to calculate the sums of squares values for all the factors.

3.4.   USE OF REPLICATES

When genuine replicates of each of the runs are made under a particular
set of experimental conditions, the variation between the responses may be
used to estimate the standard deviation of any of the runs and therefore the
standard deviation of the effects. These replicated responses will have been
obtained from exactly the same treatment combinations and should differ
only as a result of random errors. To ensure this you have to randomize the
run order and follow exactly the same procedure when setting up each of
the runs.

Suppose now you have two replicates at each of the eight conditions of
the 2^3 HPLC factorial design. You can calculate a pooled estimate of run
variance (s_i^2) using the deviations between the replicate responses. Thus

$$s_i^2 = d_i^2/2 \text{ with one degree of freedom } (d\,f\,.)$$

where d_i is the difference between the duplicate observations. You then
average the individual run variances to obtain a pooled variance estimate
(s^2) in which each separate estimate supplies one $d.f$. The duplicate
responses for the HPLC experiment and the method by which the pooled
variance estimate is obtained are given below:

Average response value	A	M	C	Results from individual runs		Difference	Estimated variance at each set of conditions
10.0	–	–	–	9.8	10.2	0.4	0.08
9.5	+	–	–	9.3	9.7	0.4	0.08
11.0	–	+	–	10.8	11.2	0.4	0.08
10.7	+	+	–	10.6	10.8	0.2	0.02
9.3	–	–	+	9.1	9.5	0.4	0.08
8.8	+	–	+	8.5	10.1	0.6	0.18
11.9	–	+	+	11.9	11.9	0.0	0.00
11.7	+	+	+	11.6	11.8	0.2	0.02
						Total =	0.54

Pooled estimate = 0.54/8 = 0.0675 with eight degrees of freedom

Since each main effect and interaction is a statistic of the form

$$\bar{y}_+ - \bar{y}_-$$

where each average contains eight observations, the variance of each effect is given by

$$V \text{ (effect)} = V(\bar{y}_+ - \bar{y}_-) = (1/8 + 1/8)s^2$$

To generalize, if N runs are used in conducting a two-level factorial or replicated factorial design;

$$V \text{ (effect)} = \frac{4}{N} s^2$$

The estimated variance of any effect in the HPLC example is thus

$$V \text{ (effect)} = \frac{4}{16} \times 0.0675 = 0.0169$$

Having calculated the variance of an effect, you may then estimate the standard error of an effect from the square root of this value. Thus:

Estimated standard error = $\sqrt{0.0169} = 0.13$

This statistic is very useful in the interpretation of the observed effects and will be used in the following section. Even though duplicate responses are available for this experiment, it is preferable in most situations to run a factorial experiment with more factors rather than replicates because many factors can influence the response and there is a limit to the time and effort you can devote to an experiment. Also later on I will introduce you to a technique which takes full advantage of all the calculated effects without the requirement for replication.

3.5.   INTERPRETATION OF EFFECTS

At this point it is perhaps useful to stop for a moment and consider why you are carrying out these calculations. Do you wish to gain an insight into the statistical significance of the effects or do you, as a scientist, merely wish to find out what size of effect you can expect when changing one or more of the factors? There is nothing stopping you from producing an analysis of variance table for every factorial design experiment you successfully complete. However, statistical significance is not the be-all and end-all of these designs and you are more likely to want some similar way of interpreting effects. Statisticians have actually been aware of this for some time and have developed some alternative methods. Some of these will be given in the following sections.

3.5.1. Comparison With a Reference Distribution

The estimated effects of the factorial analysis are given below along with the standard error of the effects calculated from the replicated runs:

Effect		Estimate	$\pm$	standard error
Main effects				
acetic acid	*A*	−0.375	$\pm$	0.13
methanol	*M*	1.925	$\pm$	0.13
citric acid	*C*	0.125	$\pm$	0.13
Two-factor interactions				
AM		0.125	$\pm$	0.13
AC		0.025	$\pm$	0.13
MC		0.825	$\pm$	0.13
Three-factor interaction				
AMC		0.025	$\pm$	0.13

Comparison of the effects with their standard errors enables us to decide which effects require interpretation and which effects may just be the result of random errors. It is quite useful to plot out these effects with a reference *t* distribution obtained from the standard error superimposed over them (Fig. 3.5a).

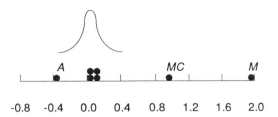

Fig. 3.5a. *Main effects and interactions in relation to a reference t distribution (eight d.f.)*

Of the main effects, factors A and M produce changes in the response which are greater than can be accounted for by residual error. Also, only one of the two-factor interactions (MC) is larger than the standard error and needs to be looked at further. An important rule about main effects is that you should only individually interpret effects if there is no evidence that the factor interacts with others. Where interactive effects seem likely from an analysis of the responses, you should discuss them jointly.

From the above table some conclusions about the influence of the factors upon the CRF may be drawn:

1. The effect of the acetic acid (A) is to reduce the CRF by approximately 0.4 irrespective of the tested levels of the other factors.

2. The effects of methanol (M) and citric acid (C) must be interpreted jointly because the MC interaction is fairly substantial. A plot of the response against citric acid concentration for the two levels of methanol substantiates this (Fig. 3.5b). If the lines connecting two levels of citric acid had been parallel then the two main effects could have been considered additive and therefore not interactive. Clearly this interaction has a great impact upon the CRF and should be borne in mind in any further studies.

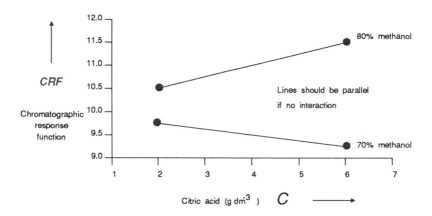

Fig. 3.5b. *Chromatographic response function (CRF) plotted against citric acid concentration for the two methanol levels*

From a practical point of view the aim of the experiment was to optimize the HPLC system. The highest *CRF* value was obtained in run 7 ($A-,M+,C+$). This may be acceptable, but the effects obtained indicate that higher *CRF* values might be obtained at an increased proportion of methanol and an increased concentration of citric acid. In Part 5 methods of optimizing such systems will be examined.

3.5.2.   Plotting Effects and Residuals on Normal Paper

In unreplicated factorial designs it is commonplace to combine the sums of squares due to the higher-order interactions and use them as the residual error sum of squares. Occasionally, however, it is likely that one of these high-order interactions is a real and meaningful effect. Using a reference distribution does not get around this problem because the standard errors have to be calculated from these high-order interactive effects. Plotting the effects on normal probability paper provides an effective way around this problem. The technique works because most of the (negligible) effects will fall on a straight line, whereas any large effects will not.

You should be fairly familiar with the normal distribution. However, to recap, it is obtained by plotting the frequency of a value against the value for a large set of data, the values of which have certain probabilities of occurring. For example, values in the extreme tails of a normal distribution have much less chance of occurring than values near the centre. In a factorial experiment large values for the effects (either positive or negative) have little chance of occurring unless they are caused by a factor or combination of factors and most values of effects will lie near zero. When a cumulative normal distribution is plotted it is sigmoid in shape. If sample distributions are now plotted this sigmoid shape makes it difficult to assess deviation from the normal. Normal probability paper helps adjust the shape of the normal curve to give a straight line when cumulative probability (P) is plotted against the value of the effect (X). Any significant effects will then show up as deviations from the line.

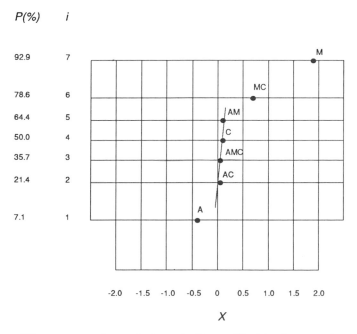

Fig. 3.5c. *Normal plot of effects, 2^2 HPLC example*

Suppose, for example, that the seven effects for the 2^3 HPLC example are plotted in this manner. In Fig. 3.5c, it is clear that some of the effects do not lie on a straight line and are of different magnitude to what may be expected when the effects are negligible. These are the A, M and MC effects which were found to be important in previous sections. How are such plots obtained? Well, to straighten out the line each of the effects is first ordered from lowest to highest as shown below.

i rank	1	2	3	4	5	6	7
effect	−0.375	0.025	0.025	0.125	0.125	0.825	1.925
identity	A	AC	AMC	C	AM	MC	M
$P = 100(i - 0.5)/T$	7.14	21.43	35.71	50	64.88	78.57	92.86

in which T is the total number of effects. If the effect is obtained at random it has an expected probability which can be calculated since you would expect the lowest of the seven effects to come in the lowest one-seventh (14.28%) of the normal distribution. This effect is then expected to lie half-way between zero and 14.28% = 7.14%. All the expected probabilities can be calculated similarly from

$$P_i = 100 \times (i - 0.5)/T$$

where P_i is the expected probability of the ith effect and i is the rank of the effect between 1 and T.

∏   Calculate the expected probability of the lowest of 10 effects.

Given the above equation, P_i is calculated as

$$P_i = 100 \times (1 - 0.5)/10 = 5\%$$

This technique is particularly applicable to screening experiments with large numbers of factors, in which the objective is to discover which effects are important.

One problem you might have come across when attempting to construct these plots is a lack of properly scaled normal probability paper. Never fear, help is at hand! Fig. 3.5d has the scale already worked out for you. You can use this to assess the position of the expected probabilities.

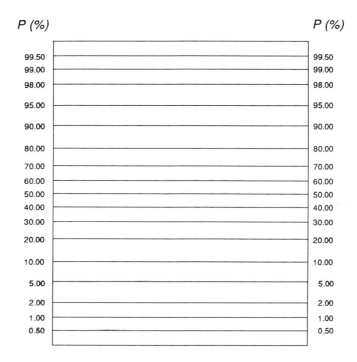

Fig. 3.5d. *Scale for producing normal plots of effects and residuals*

Let us now return to the effects which we determined for the HPLC example. Please remember that when plotted on normal paper only the *A*, *M* and *MC* effects seemed to be important. These effects spanned from −1 to +1 for each factor. If you wish to predict what the response for any of the runs might be, you have to apply the appropriate signs and divide the effects by two. Thus from Fig. 3.3e

$$\hat{y} = 10.363 + (-0.375/2)x_A + (1.925/2)x_M + (0.825/2)x_{MC}$$

where x_A, x_M and x_{MC} take the contrast signs for the particular runs.

∏   What is the predicted response for run 1?

I calculated that it should be 10.0005, since each coefficient for the main effects takes a − sign and the two-factor *MC* coefficient takes a + sign in the following equation:

$$\hat{y}_1 = 10.363 - (-0.1875) - (0.9625) + (0.4125) = 10.0005$$

The other predicted responses can be calculated in the same way for the remaining runs by applying the coefficients;

$$\hat{y}_2 = 10.363 + (-0.1875) - (0.9625) - (0.4125) = 8.8005$$

$$\hat{y}_3 = 10.363 - (-0.1875) + (0.9625) - (0.4125) = 11.1005$$

$$\hat{y}_4 = 10.363 + (-0.1875) + (0.9625) + (0.4125) = 11.5505$$

$$\hat{y}_5 = 10.363 - (-0.1875) - (0.9625) + (0.4125) = 10.0005$$

$$\hat{y}_6 = 10.363 + (-0.1875) - (0.9625) - (0.4125) = 8.8005$$

$$\hat{y}_7 = 10.363 - (-0.1875) + (0.9625) - (0.4125) = 11.1005$$

$$\hat{y}_8 = 10.363 + (-0.1875) + (0.9625) + (0.4125) = 11.5505$$

The differences between the predicted responses and the observed responses are the residuals:

Run	1	2	3	4
y	10.0	9.5	11.0	10.7
$\hat{y}$	10.0005	8.8005	11.1005	11.5505
$y - \hat{y}$	−0.0005	0.6995	−0.1005	−0.8505

Run	5	6	7	8
y	9.3	8.8	11.9	11.7
y	10.0005	8.8005	11.1005	11.5505
$y - \hat{y}$	−0.7005	−0.0005	0.7995	0.1495

Like the effects these residuals should vary about zero in a normal manner and can also be plotted on normal probability paper to see whether the variation in the data has been reasonably well accounted for in the *A*, *M* and *MC* effects. This is shown in Fig. 3.5e. Unfortunately, these residuals for this example do not lie on a straight line. This is always the difficulty with a small number of points (which is obviously a function of the size of the experiment) and we cannot expect a perfect fit for the residuals.

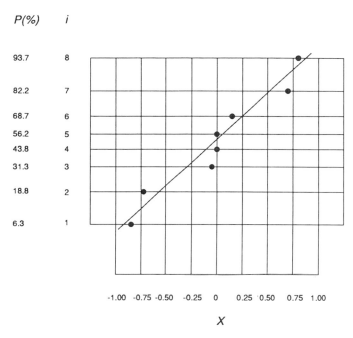

Fig. 3.5e. *Normal plot of residuals, 2^2 HPLC example*

SAQ 3.5

Again using the plating example, extract the calculated effects from one of the previous SAQ responses and plot them on the normal probability paper in Fig. 3.5f. dECIDE which effects cannot be accounted for by random variations in the response. Calculate the residuals, plot them in order on Fig. 3.5g and decide whether the predicted effects sufficiently account for the observed responses.

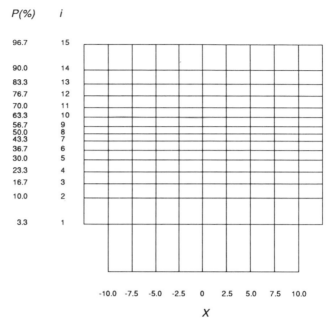

Incomplete Fig. 3.5f. *Normal plot of effects, 2^4 factorial design, plating example*

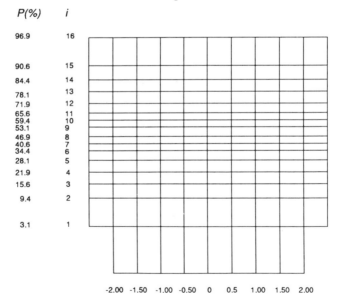

Incomplete Fig. 3.5g. *Normal plot of residuals, 2^4 factorial, plating example*

Learning Objectives

After studying the material in Part 3, you should now be able to:

- understand the potential impact upon the experimental response of interactions between variables;

- design complete factorial design experiments for any number of variables;

- determine and interpret the main effects and interactions of variables;

- apply columns of contrast coefficients and Yates' algorithm as short cut methods of calculating these effects;

- derive analysis of variance tables using higher-order interactions as estimates of residual error;

- compare the effects with a suitable reference distribution when available;

- apply the method of plotting the effects and residuals on normal probability paper.

SAQs AND RESPONSES FOR PART THREE

SAQ 3.1a	Match the words on the left-hand side to the most appropriate phrase on the right-hand side.

(a) factor	1) an observed numerical result
(b) levels	2) when the effects of two or more factors are not additive
(c) response	3) a combination of factor levels used in an experiment
(d) interaction	4) different values for a factor
(e) treatment	5) a variable believed to affect the outcome of an experiment

Response

I trust you managed most of these quite easily. However, whether you did
or not the answers are as follows:

(a) and 5) A factor is a variable believed to affect the outcome of an
experiment. (The terms variable and factor will be used almost
synonymously throughout this unit).

(b) and 4) Levels are simply the different values for a factor. This applies
equally to both quantitative factors where it is easy to visualize
changes in levels, and qualitative variables where a factor may
simply be present or absent.

(c) and 1) A response is an observed result obtained for each experimental
run. For most experimental purposes the response will be a
numerical value, although this is not always possible.

(d) and 2) An interaction occurs when the effects of two or more factors
are not additive.

(e) and 3) A treatment is the combination of factor levels used in an
experimental run. For example, two factors, each at two levels,
would have four different treatments.

****

SAQ 3.1b
> In an inter-laboratory study several samples are
> to be analysed by different methods at different
> temperatures. How many different effects should you
> take into account?

Response

I hope you are not shocked to learn that the answer is seven. You probably
realised that there are three main effects, one due to the effect of the sample,

another due to the different methods being employed and another due to variation in the temperature at which the analysis is to be carried out. You might also have realised that there are three possible interactions between pairs of these factors, namely sample and method, sample and temperature, and method and temperature. However, it is also possible to have a complex interaction between all three factors. This is termed the sample–method–temperature three-factor interaction.

If you were unsure about answering this question, you might find it instructive to re-read Section 3.1 of this Unit.

****

SAQ 3.1c	For an unreplicated 3^4 design
	1. How many factors (variables) are to be studied?
	2. How many levels are these factors to be studied at?
	3. What is the total number of runs required?

Response

1. The correct answer here is 4 because this is the power term in the specified design. If the design was mixed, that is it had one factor at one set of levels and other factors at other levels, then it would be the sum of the power terms specified.

2. This design will estimate the effects of each factor at three levels, so allowing estimates of the linear and quadratic effects of each of the factors.

3. The total number of runs can be calculated from the design specification;

$3^4 = 3 \times 3 \times 3 \times 3 = 81$ runs.

****

SAQ 3.2   Given below is a 2^4 factorial design to examine the effects of four factors, current (C), ratio of time of plating ON to OFF (R), strength of the acid solution (A), and frequency of the plating signal (F), upon an electroplating procedure. Give two alternatives to the notation used.

Factor

C	R	A	F
−	−	−	−
+	−	−	−
−	+	−	−
+	+	−	−
−	−	+	−
+	−	+	−
−	+	+	−
+	+	+	−
−	−	−	+
+	−	−	+
−	+	−	+
+	+	−	+
−	−	+	+
+	−	+	+
−	+	+	+
+	+	+	+

Response

Given that the design is one of four variables (factors), each studied at two levels, you could have used the lower case letters where the factors are at the higher levels, that is where there are + signs. The run with all the factors at their lower levels (run 1) is designated by the number (*1*).

Alternatively, you could have written the design matrix with the numbers 0 and 1 under the factors to represent the lower and higher levels respectively. This notation is essentially very similar to that given above!

The completed table with both alternative notations could look something like the following:

Factor					Factor			
C	R	A	F		C	R	A	F
−	−	−	−	*(1)*	0	0	0	0
+	−	−	−	*c*	1	0	0	0
−	+	−	−	*r*	0	1	0	0
+	+	−	−	*cr*	1	1	0	0
−	−	+	−	*a*	0	0	1	0
+	−	+	−	*ca*	1	0	1	0
−	+	+	−	*ra*	0	1	1	0
+	+	+	−	*cra*	1	1	1	0
−	−	−	+	*f*	0	0	0	1
+	−	−	+	*cf*	1	0	0	1
−	+	−	+	*rf*	0	1	0	1
+	+	−	+	*crf*	1	1	0	1
−	−	+	+	*af*	0	0	1	1
+	−	+	+	*caf*	1	0	1	1
−	+	+	+	*raf*	0	1	1	1
+	+	+	+	*craf*	1	1	1	1

****

SAQ 3.3a

Complete the following paragraph by inserting the most appropriate word or phrase, chosen from the list given below, into the blank spaces.

Two-level factorial designs estimate not only ....
effects but also .... effects with maximum
.... The main effects may be estimated as
the difference between two ...., one with the
.... at its + level, the other at its − level. Two-
factor interactive effects may be estimated from half
the .... between an average main effect at the
.... level of another factor subtracted from the
average effect at the ...... level. The .... of the
effects may be judged from an estimate of the ....
of an effect obtained by genuine replication when
available.

HIGHER	AVERAGES
TWO-LEVEL	MAIN
INTERACTIVE	FACTOR
ACCURACY	RESULTS
VARIANCE	MAGNITUDE
PRECISION	SIGNIFICANCE
STANDARD ERROR	DIFFERENCE
LOWER	

Response

Two-level factorial designs estimate not only MAIN effects but also INTERACTIVE effects with maximum PRECISION. The main effects may be estimated as the difference between two AVERAGES, one with the FACTOR at its + level, the other at its − level. Two-factor interactive effects may be estimated from half the DIFFERENCE between an average main effect at the LOWER level of another factor subtracted from the average effect at the HIGHER level. The SIGNIFICANCE of the effects may be judged from an estimate of the STANDARD ERROR of an effect obtained by genuine replication when available.

****

SAQ 3.3b

Given below are the results of a 2^4 factorial experiment to examine the effects of four factors, current (C), ratio of time of plating ON to OFF (R), strength of the acid solution (A), and frequency of the plating signal (F), upon an electroplating procedure. Each of the factors was investigated at two levels, − and +. Use columns of contrast coefficients to calculate the main effects and interactions of the factors. The response was the hardness of the plate.

| Factor | | | | |
C	R	A	F	Hardness
−	−	−	−	40
+	−	−	−	42
−	+	−	−	38
+	+	−	−	41
−	−	+	−	52
+	−	+	−	54
−	+	+	−	48
+	+	+	−	49
−	−	−	+	46
+	−	−	+	48
−	+	−	+	44
+	+	−	+	47
−	−	+	+	58
+	−	+	+	60
−	+	+	+	54
+	+	+	+	55

Response

The effects I calculated are given below:

Effect	Estimate
Average	48.5
Main effects	
current C	2.0
ratio of time ON/OFF R	-3.0
acid strength A	10.5
frequency F	6.0
Two-factor interactions	
CR	0.0
CA	-0.5
CF	0.0
RA	-1.5
RF	0.0
AF	0.0
Three-factor interactions	
CRA	-0.5
CRF	0.0
CAF	0.0
RAF	0.0
Four-factor interaction	
$CRAF$	0.0

To calculate all the main effects and interactions I first multiplied the signs in the columns of the original design matrix to produce the columns corresponding to the interactive effects. You should have obtained a series of columns corresponding exactly to the one I produced below. Note that each column has eight − signs and eight + signs, so that each effect is calculated as a contrast between two sets of eight experiments.

The + sign for the first run under RA was obtained as the product of the − signs under R and A, the + sign of the second run was the product of the − sign under R and the − sign under A and so on. The columns of contrast coefficients for each of the interactions was calculated in the same manner.

C	R	A	F	CR	CA	CF	RA	RF	AF	CRA	CRF	CAF	RAF	CRAF
−	−	−	−	+	+	+	+	+	+	−	−	−	−	+
+	−	−	−	−	−	−	+	+	+	+	+	+	−	−
−	+	−	−	−	+	+	−	−	+	+	+	−	+	−
+	+	−	−	+	−	−	−	−	+	−	−	+	+	+
−	−	+	−	+	−	+	−	+	−	+	−	+	+	−
+	−	+	−	−	+	−	−	+	−	−	+	−	+	+
−	+	+	−	−	−	+	+	−	−	−	+	+	−	+
+	+	+	−	+	+	−	+	−	−	+	−	−	−	−
−	−	−	+	+	+	−	+	−	−	−	+	+	+	−
+	−	−	+	−	−	+	+	−	−	+	−	−	+	+
−	+	−	+	−	+	−	−	+	−	+	−	+	−	+
+	+	−	+	+	−	+	−	+	−	−	+	−	−	−
−	−	+	+	+	−	−	−	−	+	+	+	−	−	+
+	−	+	+	−	+	+	−	−	+	−	−	+	−	−
−	+	+	+	−	−	−	+	+	+	−	−	−	+	−
+	+	+	+	+	+	+	+	+	+	+	+	+	+	+

I then applied the coefficients to the results of the experiments and divided the result by 8, since this is the number of comparisons made in the 16 experiments. Thus the interaction between factors R and A is given by

$$RA = \frac{\begin{array}{l}(+40 + 42 - 38 - 41 - 52 - 54 + 48 + 49 + 46 \\ \quad + 48 - 44 - 47 - 58 - 60 + 54 + 55)\end{array}}{8}$$

$$= -1.5$$

If you did not understand how I derived the contrasts and used these to calculate the effects then re-read Section 3.3.3.

$$**********************************$$

SAQ 3.3c

In an investigation into extraction of nitrate-nitrogen from air-dried soil, three quantitative variables were investigated at two levels. These were the amount of oxidized activated charcoal (A) added to the extracting solution to remove organic interferences, the strength of the $CaSO_4$ extracting solution (C), and the time that the soil was shaken with the solution (T). The aim of the investigation was to optimize the extraction procedure. The levels of the variables are given below:

		−	+
Activated charcoal (g)	A	0.5	1.0
$CaSO_4$ (%)	C	0.1	0.2
Time (minutes)	T	30	60

The concentrations of nitrate-nitrogen were determined by ultra-violet spectrophotometry and compared with concentrations determined by a standard technique. The results given below are the amounts recovered (expressed as the percent of the known nitrate concentration).

Run	A	C	T	Percent total (y)
1	−	−	−	68
2	+	−	−	65
3	−	+	−	75
4	+	+	−	72
5	−	−	+	90
6	+	−	+	89
7	−	+	+	99
8	+	+	+	95

Calculate and interpret the main effects of the factors and all the possible two- and three-factor interactions.

Response

The effects I calculated are given below:

Effect	Estimate
Average	81.63
Main effects	
activated charcoal A	−2.75
$CaSO_4$ C	7.25
time T	23.25
Two-factor interactions	
AC	−0.75
AT	0.25
CT	0.25
Three-factor interaction	
ACT	−0.75

To obtain the main effects I averaged the experimental responses at the lower levels (−) and subtracted this figure from the average at the higher levels (+). Thus, I averaged the responses from runs 1, 3, 5 and 7 (83%) and runs 2, 4, 6 and 8 (80.25%). The difference between these is −2.75%. The negative effect indicates that lower recoveries were obtained when 1 gram of activated charcoal was added to the extracting solution; and may have been due to increased adsorption of nitrate by the charcoal.

The effect of changing the strength of the extracting solution (C) was similarly calculated by subtracting the average of the results from runs 1, 2, 5 and 6 (78%) from the average of runs 3, 4, 7 and 8 (85.25%). The same procedure was followed for the time effect (T).

I calculated the two-factor interactions by considering that there were two estimates of each main effect at two levels of the other factors. Thus, there are two estimates of A, one at the lower level of C, the other at the higher level. By convention half the difference between these estimates is called the charcoal by strength of extracting solution interaction (AC).

These are given below:

CaSO$_4$	Average activated charcoal effect
(+) 0.2% −3.5 $= \dfrac{(-3)+(-4)}{2} = (y_4 - y_3 + y_8 - y_7)/2$	
(−) 0.1% −2.0 $= \dfrac{(-3)+(-1)}{2} = (y_2 - y_1 + y_6 - y_5)/2$	
difference −1.5	
AC interaction $= \dfrac{-1.5}{2} = -0.75$	

I could also have examined the consistency of the C effect at the two levels of activated charcoal. I calculated the other two-factor interactions between time and charcoal (AT) and between time and solution strength (CT) in a similar manner.

Activated charcoal	Average time effect	CaSO$_4$	Average time effect
(+) 1 g	23.5	(+) 0.2%	23.5
(−) 0.5 g	23.0	(−) 0.1%	23.0
	difference 0.5		difference 0.5
AT interaction $= \dfrac{0.5}{2} = 0.25$		CT interaction $= \dfrac{0.5}{2} = 0.25$	

Obviously your calculations may have involved calculating the consistency of the other effect in each pair of factors but you should have obtained the same results.

To calculate the three-factor interaction, I considered that there were two estimates of the $A \times C$ two-factor interaction available, one for each level of time (T).

AC interaction with time (+)

$$= \frac{(y_8 - y_7) - (y_6 - y_5)}{2} = \frac{(95 - 99) - (89 - 90)}{2} = \frac{(-4) - (-1)}{2}$$

$$= -1.5$$

AC interaction with time (−)

$$= \frac{(y_4 - y_3) - (y_2 - y_1)}{2} = \frac{(72 - 75) - (65 - 68)}{2} = \frac{(-3) - (-3)}{2}$$

$$= 0$$

Half the difference between these two estimates is defined as the three-factor interaction of activated charcoal, $CaSO_4$ and time, denoted as ACT. Thus,

$$ACT \text{ interaction } = \frac{-1.5 - 0}{2} = -0.75$$

This was how I calculated the three-factor interaction. You could, however, have calculated it by considering the consistency of any two-factor interaction at the two levels of the third factor. These results show that none of the interactions is particularly large in comparison to the main effects. Thus the main effects of the factors may be quoted without reference to the levels of the other factors.

****

SAQ 3.3d	Continuing with the plating example introduced in the previous SAQ, use Yates' method to calculate the sums of squares values for all the factors.

Response

OK, you were given the responses in the standard order and should therefore have had no problem in starting the calculations. I trust you realised you needed 4 columns of figures because it is a 2^4 factorial design. The important point here is that you are careful about the additions and subtractions. Personally I find it a bit difficult to keep concentrating while performing all the additions and subtractions. However, if you managed to get through to column (4) correctly you should then have obtained the estimates by dividing by the appropriate number. The first number is divided by 16 to give the mean response, whereas all the others are divided by 8. You should already have these values from the previous SAQ and the two sets should agree. Here it is the sums of squares that you are after. These you should have obtained by squaring the values in column (4) and dividing by 16 (all except the first value which is equivalent to the correction for the mean). The 15 main effects and interactions that these correspond to come from + signs given in the design matrix in standard order. Thus the penultimate row has + signs under R, A and F and is the RAF sum of squares.

Factor

C	R	A	F	y	(1)	(2)	(3)	(4)	Effect	Divisor	Estimate	SS
−	−	−	−	40	82	161	364	776	Mean	16	48.5	—
+	−	−	−	42	79	203	412	16	C	8	2.0	16
−	+	−	−	38	106	185	8	−24	R	8	−3.0	36
+	+	−	−	41	97	227	8	0	CR	8	0	0
−	−	+	−	52	94	5	−12	84	A	8	10.5	441
+	−	+	−	54	91	3	−12	−4	CA	8	−0.5	1
−	+	+	−	48	118	5	0	−12	RA	8	−1.5	9
+	+	+	−	49	109	3	0	−4	CRA	8	−0.5	1
−	−	−	+	46	2	−3	42	48	F	8	6.0	144
+	−	−	+	48	3	−9	42	0	CF	8	0	0
−	+	−	+	44	2	−3	−2	0	RF	8	0	0
+	+	−	+	47	1	−9	−2	0	CRF	8	0	0
−	−	+	+	58	2	1	−6	0	AF	8	0	0
+	−	+	+	60	3	−1	−6	0	CAF	8	0	0
−	+	+	+	54	2	1	−2	0	RAF	8	0	0
+	+	+	+	55	1	−1	−2	0	CRAF	8	0	0

****

SAQ 3.5

Again using the plating example, extract the calculated effects from one of the previous SAQ responses and plot them on the normal probability paper in Fig. 3.5f. Decide which effects cannot be accounted for by random variations in the response. Calculate the residuals, plot them in order on Fig. 3.5g and decide whether the predicted effects sufficiently account for the observed responses.

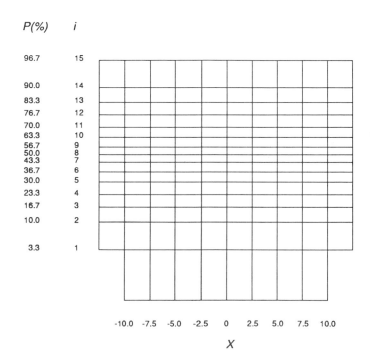

Incomplete Fig. 3.5f. *Normal plot of effects, 2^4 factorial design, plating example*

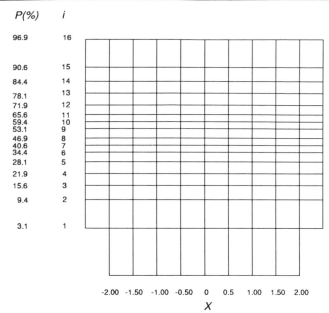

Incomplete Fig. 3.5g. *Normal plot of residuals, 2^4 factorial, plating example*

Response

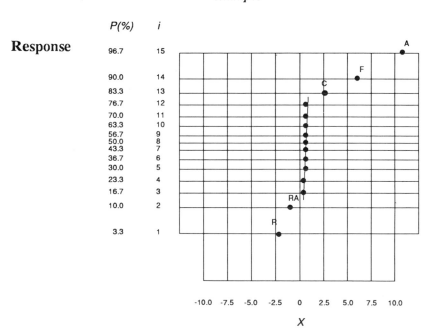

Completed Fig. 3.5f. *Normal plot of effects, 2^4 factorial design, plating example*

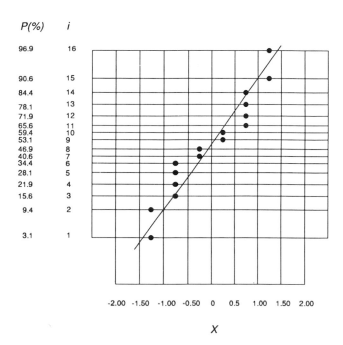

Completed Figure 3.5g. *Normal plot of residuals, 2^4 factorial design, plating example*

Your completed figure should match mine fairly closely, especially as I put on the probability scale for you. All you had to do was to rank the effects and plot the values on the X axis. Examination of this figure would seem to indicate that very few of the effects actually deviate from the straight line. It is interesting to see that the large effects are all due to factors themselves with very few important interactions between the factors. There is some slight degree of interaction between R and A and it would be advisable for you to look at this more closely, possibly with a follow-up experiment. Assuming that the main effects and RA two-factor interaction are the only important effects it is now possible to calculate the predicted responses for all sixteen experimental runs. Remembering to divide each effect by two because the levels used actually span the factor space from -1 to $+1$, the predicted response for the first run is

$$\hat{y} = 48.5 + (2/2)x_C + (-3/2)x_R + (10.5/2)x_A + (6/2)x_F + (-1.5)x_{RA}$$

where x_C, x_R etc. are the contrasts. Therefore, for the first run:

$$\hat{y}_1 = 48.5 - (1.0) - (-1.5) - (5.25) - (3.0) + (-1.5) = 39.25$$

A complete list of predicted responses can be obtained in the same way for all the runs and the residuals obtained by subtraction from the observed responses:

Run	y	$\hat{y}$	Residual
1	40	39.25	0.75
2	42	41.25	0.75
3	38	39.25	-1.25
4	41	41.25	-0.25
5	52	52.75	-0.75
6	54	54.75	-0.75
7	48	46.75	1.25
8	49	48.75	0.25
9	46	45.25	0.75
10	48	47.25	0.75
11	44	45.25	-1.25
12	47	47.25	-0.25
13	58	58.75	-0.75
14	60	60.75	-0.75
15	54	52.75	1.25
16	55	54.75	0.25

Most of the residuals seem to be fairly small in comparison to the effects. You should have put them in order and completed the plot of the residuals (Fig. 3.5g) which should agree with the one I have prepared above in which the residuals more or less fit to the line.

If you have a problem with these plots it may not be entirely due to a lack of understanding at this level, and may stretch back to a lack of understanding about normal distributions which you can clear up by looking at one of the basic texts on statistics. Otherwise re-read the section from the beginning and have another go at the problem.

****

4. Fractional Factorials

4.1. REDUNDANCY OF ESTIMATES

One of the main disadvantages of full or complete factorial experimental designs is that the number of experimental runs required to estimate all the main effects and interactions increases rapidly as the number of factors increases. In Part 3 it was shown that two-level factorial designs in f factors require 2^f runs. Therefore, if six factors are to be examined, 64 experimental runs are required in which each run represents a different treatment. This experiment can estimate up to 64 different statistics including one average, six main effects, 15 two-factor interactions, 20 three-factor interasactions, 15 four-factor interactions, six five-factor interactions and one six-factor interaction.

∏   How many different statistics can be obtained from a 2^7 design?

The answer I hope you got was 128 since the number of runs in a factorial experiment is the number of levels raised to the power of the number of factors, or more simply, you could have doubled the number for a 2^6 experiment (= 64 × 2 = 128). You will also have realised that such a design will estimate the main effects of seven variables. However, you probably won't have been able to work out the partitioning of the interactions amongst the remaining effects. A quick method for doing this is supplied by the following equation:

$$\text{Number of } p\text{-factor interactions in a } 2^f \text{ design} \ = \ \frac{f!/(f-p)!}{p!}$$

where ! indicates the factorial of the number (for example, the factorial of 3 is given by 3! = 3 × 2 × 1 = 6), f is the number of factors and p is the level of the effect you wish to find the number of interactions for. Applying this to calculate the number of three-factor interactions in a 2^7 design we get

$$\text{No. of three-factor interactions} = \frac{7!/(7-3)!}{3!} = \frac{7 \times 6 \times 5}{3 \times 2 \times 1} = 35$$

∏   How many four-factor interactions are estimated in a complete 2^6 factorial design?

Applying the above formula:

$$\text{No. of four-factor interactions} = \frac{6!/(6-4)!}{4!} = \frac{6 \times 5 \times 4 \times 3}{4 \times 3 \times 2 \times 1} = 15$$

Of course you could apply this formula to all 2^f designs but it would get a bit laborious. Therefore, I have supplied a complete table of the number of main effects and interactions for values of f up to 8:

Effect	$f = 2$	$f = 3$	$f = 4$	$f = 5$	$f = 6$	$f = 7$	$f = 8$
Average	1	1	1	1	1	1	1
Main	2	3	4	5	6	7	8
2-factor	1	3	6	10	15	21	28
3-factor		1	4	10	20	35	56
4-factor			1	5	15	35	70
5-factor				1	6	21	56
6-factor					1	7	28
7-factor						1	8
8-factor							1

From the tabulated values it can be seen that when f is large the number of complex interactions involving three or more factors becomes extremely large. This raises an important point. Even where the effects of higher-order interactions can be estimated, it does not imply that they are of significance! More often these higher-order effects are small in comparison to main effects and two-factor interactions. For most experimental situations main effects tend to be larger than two-factor interactions, which in turn tend to be larger than three-factor interactions and so on, so that at some point higher-order interactions can be regarded as negligible or redundant. Also, when there is a large number of factors in a factorial design, it often happens that very few are important. Fractional factorial designs exploit both these aspects of factorial designs by disregarding the possible importance of high-order interactions, and use only a fraction of the experimental runs required for the complete design.

The main advantage of fractional factorial designs is that they can be used to investigate the effects of a large number of factors in very few runs. This is something you might wish to do in the early stages in an investigation when little is known about the system with which you are dealing. It is therefore desirable to run an experiment which is capable of telling you which factors are important. Such a design should have the properties of large f and small N. Should some of the factors subsequently appear to be important, other runs or whole new experiments with similar designs may be conducted to make sure of this. Alternatively, other types of experiment such as response surface designs or sequential techniques such as steepest ascent or simplex designs can be applied to optimize the response by adjusting the combination of levels of the important factors.

4.2. HALF-FRACTION FACTORIAL DESIGNS

To illustrate the point about using a small number of runs to determine whether lots and lots of factors are important, an example is taken from a study of the factors involved in the response of an automatic chemical analyzer. This automatic chemical analyzer is used to determine the activity of a serum enzyme, alkaline phosphatase. The aims of this preliminary study are to determine which of the factors influence the response (phosphatase enzyme activity), to decide whether interactions between the factors are important, and if so, to estimate the magnitudes of these effects quantitatively.

From previous experience it is believed that the factors which influence the phosphatase activity are zinc sulphate (Z), magnesium sulphate (M), pH (P), disodium p-nitrophenyl phosphate (D) and 2-amino-2-methyl-1-propanol (A). A complete 2^5 design experiment was carried out with the factors at the following levels:

Factor	Units	Level	
		$-$	$+$
Z	μmol dm^{-3}	40	80
M	μmol dm^{-3}	1.50	2.50
P	dimensionless	10.00	10.70
D	mmol dm^{-3}	10	20
A	mol dm^{-3}	0.20	0.60

The 2^5 design matrix and the responses (enzyme activity) obtained are given below:

Run	Z	M	P	D	A	Enzyme activity (IU)
1	−	−	−	−	−	109
2*	+	−	−	−	−	113
3*	−	+	−	−	−	103
4	+	+	−	−	−	113
5*	−	−	+	−	−	103
6	+	−	+	−	−	104
7	−	+	+	−	−	106
8*	+	+	+	−	−	123
9*	−	−	−	+	−	119
10	+	−	−	+	−	146
11	−	+	−	+	−	111
12*	+	+	−	+	−	143
13	−	−	+	+	−	116
14*	+	−	+	+	−	145
15*	−	+	+	+	−	110
16	+	+	+	+	−	148
17*	−	−	−	−	+	106
18	+	−	−	−	+	120
19	−	+	−	−	+	113
20*	+	+	−	−	+	115
21	−	−	+	−	+	109
22*	+	−	+	−	+	117
23*	−	+	+	−	+	105
24	+	+	+	−	+	115
25	−	−	−	+	+	96
26*	+	−	−	+	+	128
27*	−	+	−	+	+	95
28	+	+	−	+	+	127
29*	−	−	+	+	+	99
30	+	−	+	+	+	131
31	−	+	+	+	+	92
32*	+	+	+	+	+	132

Of course you could now estimate all the main effects and interactions up to order 5 if you desired, by applying the methods given in Part 3. Since there are 31 of these (32 including the average) I have saved you the bother and they are given below:

Effect	Estimate	Effect	Estimate
Average	116.00	*ZMP*	2.25
		ZMD	0.63
Z	20.50	*ZMA*	−2.38
M	−0.63	*ZPD*	0.63
P	−0.13	*ZPA*	−0.13
D	10.25	*ZDA*	0.50
A	−7.00	*MPD*	−1.00
		MPA	−3.00
ZM	2.13	*MDA*	1.63
ZP	1.38	*PDA*	0.88
ZD	12.25		
ZA	0.75	*ZMPD*	−0.75
MP	1.50	*ZMPA*	0.50
MD	−2.13	*ZMDA*	1.63
MA	−0.88	*ZPDA*	0.13
PD	1.13	*MPDA*	1.50
PA	0.13		
DA	−10.25	*ZMPDA*	0.00

As there was no replication of the treatments I have applied the method of plotting the effects on normal probability paper to decide which effects, if any, are important (Fig. 4.2a).

From the normal plot it would appear that only five of the estimated effects are appreciably important. These five effects include three main effects (*Z*, *D* and *A*) and two two-factor interactions (*ZD* and *DA*). All the other interactions were found to have little influence upon the response. Obviously if I had known in advance that so few effects were going to be important I might have chosen a 2^3 design in factors *Z*, *D* and *A*, requiring only eight experimental runs, thus saving myself a lot of time and effort. The problem is that I could not have known this without carrying out some experiments. What I therefore intend to do is to show you how to apply a class of design which uses fewer runs but gives you almost as much information.

Suppose that instead of carrying out 32 runs I decided to carry out runs based on the 16 treatment combinations marked with asterisks in the original design matrix and then calculated the main effects and two-factor interactions based on these responses. You can calculate these effects if

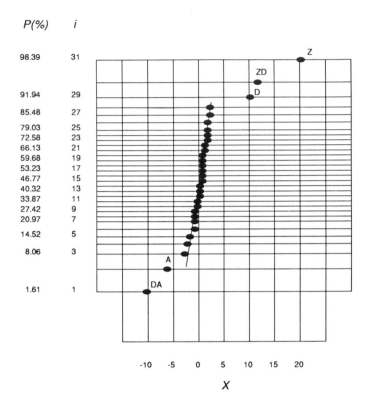

Fig. 4.2a. *Normal plot of effects from a 2^5 factorial design, alkaline phosphatase example*

you wish but they are given below anyway (I hope you agree with my estimates!):

Effect	Estimate	Effect	Estimate
Average	116.00	ZM	3.00
		ZP	3.00
Z	22.00	ZD	9.25
M	−0.50	ZA	−0.25
P	1.50	MP	2.00
D	10.75	MD	−2.25
A	−7.75	MA	−0.25
		PD	−1.25
		PA	0.75
		DA	−8.00

The main effects of Z, D and A are again large and the two-factor interactions ZD and DA also appear to be important. A normal plot of these effects (Fig. 4.2b) seems to show that, of the main effects and two-factor interactions, these five effects are the only ones to influence the phosphatase activity to any great extent. However, only half the previous number of runs were required. "Miraculous!" I hear you say. "No!" I say. I merely chose the treatment combinations for the factors carefully. You will of course have noticed that I haven't attempted to calculate any interactions involving three or more factors and I shall explain why later!

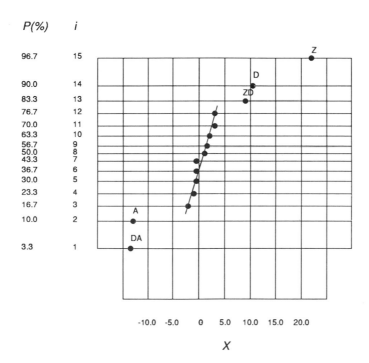

Fig. 4.2b. *Normal plot of effects for a 2^{5-1} fractional factorial design, alkaline phosphatase example*

One way I could have attempted to estimate the effects in the sixteen runs would have been to use columns of contrast coefficients (Section 3.3.3). Obviously the design matrix supplies the contrasts for the main effects directly, but not the interactions (which are obtained by multiplying the main effect signs). Instead of writing out all the contrasts, I've written out columns of contrast coefficients for just a few of the main effects and two-factor interactions which were estimated, and also some of the interactions which were not estimated. With the aid of these contrasts I hope to show you what has happened to the higher-order interactions.

				Effect			
Z	MPDA	A	ZMPD	ZM	PDA	ZA	MPD
−	−	+	+	+	+	−	−
+	+	−	−	−	−	−	−
−	−	−	−	−	−	+	+
+	+	+	+	+	+	+	+
−	−	−	−	+	+	+	+
+	+	+	+	−	−	+	+
−	−	+	+	−	−	−	−
+	+	−	−	+	+	−	−
−	−	−	−	+	+	+	+
+	+	+	+	−	−	+	+
−	−	+	+	−	−	−	−
+	+	−	−	+	+	−	−
−	−	+	+	+	+	−	−
+	+	−	−	−	−	−	−
−	−	−	−	−	−	+	+
+	+	+	+	+	+	+	+

There are some similarities between the main effect and four-factor interaction columns. Notice, for example, that the column of signs for Z is in fact identical to the column of signs for the $MPDA$ four-factor interaction and that the signs for M are the same as those of $ZPDA$. Clearly these main effect estimates will be the same as the estimates of the four-factor interactions because the signs are identical. Similarly, the column of signs for the ZM two-factor interaction has the same signs as the column for the PDA three-factor interaction, and ZA has the same signs as MPD. Consequently the estimates of these pairs of effects will be the same. When such a situation occurs the effects are said to be *aliases* of one another, because they are indistinguishable, or to put it in statistical jargon they are *confounded*.

Although I haven't provided all the columns of signs above, each main effect and two-factor interaction has an alias and this is the major difference between the fractional and complete factorial designs. In the complete design each column of signs is unique, whereas we have just established that each effect in the fractional design has an alias. Instead of being able to estimate 32 effects, only 16 are now available, each of which is actually a sum of the two effects. This includes the average response which is aliased with the five-factor effect. This loss of clear-cut estimations is the price that has been paid for using fewer experimental runs. A more accurate picture of the effects is obtained if each main effect and two-factor interaction is expressed alongside its alias.

Effect	Estimate
Average + *ZMPDA*	116.00
Z + *MPDA*	22.00
M + *ZMDA*	−0.50
P + *ZMDA*	1.50
D + *ZMPA*	10.75
A + *ZMPD*	−7.75
ZM + *PDA*	3.00
ZP + *MDA*	3.00
ZD + *MPA*	9.25
ZA + *MPD*	−0.25
MP + *ZDA*	2.00
MD + *ZPA*	−2.25
MA + *ZPD*	−0.25
PD + *ZMA*	−1.25
PA + *ZMD*	0.75
DA + *ZMP*	−8.00

The justification for combining the effects is that it is unlikely that the higher-order alias is going to be important, a risk which is taken in designing the fractional experiment.

Of course I used 16 runs for five factors in the fractional design whereas the full 2^5 design uses 32. This fractional factorial design is known as a half-fraction or half-replicate because it uses half the runs of the full factorial. It is designated as

$$2^{5-1}$$

since it is used to investigate five factors in $2^{5-1} = 2^4 = 16$ runs (half the original 32 runs). A vast array of fractions of complete factorial designs is available to help the experimenter investigate a large number of factors.

Π   How many factors are investigated in a 2^{5-2} fractional factorial design? How many runs are required?

A 2^{5-2} design can estimate the effects of five factors as indicated by the first superscript. The number of runs required is

$$2^{5-2} = 2^3 = 8$$

This experiment is a quarter-fraction since the full factorial design in five variables requires 32 runs. The fraction is given by the ratio of the runs in both designs. Thus

$$8/32 = 1/4$$

SAQ 4.2

What fraction of a complete two-level factorial design is necessary to investigate the effects of eight factors in 16 runs?

A: 1/2   B: 1/10

C: 1/8   D: 1/16

Responses to SAQs in Part 4 begin on page 177.

So far I haven't explained how I selected the 16 treatments for the alkaline phosphatase experiment which resulted in the pattern of aliases that I wanted for the half-fraction. The answer lies in the aliasing of the effects. Remember that the main effects were aliased with four-factor interactions and the two-factor interactions were aliased with three-factor interactions. This was deliberate because I wanted the best possible estimates of the main effects. These were confounded with the four-factor interactions which were the least likely to be important. I actually wrote down a complete design matrix for the first four factors Z, M, P and D and worked out the signs for the $ZMPD$ four-factor interaction. I then used these signs to define the levels of the fifth factor, A. Wherever $ZMPD$ was +, as in the first run, the higher level of factor A was applied, and wherever $ZMPD$ was −, as in the second run, factor A was applied at its lower level. Consequently when the effect of A was estimated it was inseparable from (or confounded by) the estimate of the *ZMPD* effect. The relationship $A = ZMPD$ used in this design is called the *generator*. As a result of generating the design in this manner all the main effects were confounded with four-factor interactions and the two-factor interactions were confounded with three-factor interactions.

I should at this point apologize for apparently losing sight of the more important implications of the effects of the five chosen different factors upon the phosphatase activity. To the clinical chemist these factors may be highly relevant. I hope that for the purposes of understanding experimental design you can treat the factors I have chosen as any five factors which have to be investigated in a minimum number of runs. You may already be thinking of situations where this type of design is relevant to you and therefore applicable to your own sets of potentially important factors.

4.3.   GENERATING FRACTIONAL DESIGNS

4.3.1.   Generating Half-fraction Factorial Designs

Matching all the columns in the design is unnecessary (it also tends to make you a bit cross-eyed) since a method has been developed by which the aliases can be found from the generator. First of all you have to multiply the effects in the generator to produce a *defining relation* or *defining contrast* (I). For this example:

$$I = ZMPD \times A = ZMPDA$$

The alias of any effect may then be obtained by multiplying the effect by I using normal algebraic rules, with an additional rule that where a term appears an even number of times in the product it disappears. Therefore, to find the alias of Z:

$$Z = Z \times ZMPDA = Z^2 MPDA = MPDA$$

∏   What is the alias of *ZM*?

Using the defining relation $I = ZMPDA$, the alias of ZM is given by:

$$ZM = ZM \times ZMPDA = Z^2 M^2 PDA = PDA$$

∏   What is the difference between the aliasing structures for this half-fraction design and one generated by letting $Z = MPDA$?

Hopefully you can see immediately that these two half-fractions are exactly the same. If you can't just work out what the defining relation is. For example, we have already said that the defining relation in the first design is $I = ZMPDA$. If the second generator is $Z = MPDA$ then the defining relation is also $Z \times MPDA = ZMPDA$. Thus the effects have exactly the same aliases in both cases.

In our example the generator $A = ZMPD$ was used to obtain a half-fraction design (16 out of the total 32 runs). It would have been possible to produce a half-fraction incorporating the other 16 runs by changing the signs of one term in the generator so that $A = -ZMPD$. Instead of being the sums of two effects the estimates will now be the differences between two effects. However, the main effect estimates of this second half-fraction should be very similar to those of the first half-fraction since the four-factor interactions are fairly small. On the other hand, the estimates of the two-factor interactions are likely to be different because some three-factor interactions in the complete design were large.

SAQ 4.3a	A combination of these two half-fractions yields the complete factorial design for five variables at two levels [True/False]?

Instead of considering the design as two half-fractions, the design may be considered as a 2^5 factorial design run in two blocks. Combining the estimates of the effects yielded by both designs will give the same estimates as the original 2^5 design run as a single block.

$A = ZMPD$ $A = -ZMPD$

Estimate of Z Effect   Estimate of Z Effect

$(Z + MPDA) = 22.00$ $(Z - MPDA) = 19.00$

$(Z + MPDA) + (Z - MPDA) = 2Z\quad\quad = 22.00 + 19.00 = 41.00$
$(Z + MPDA) - (Z - MPDA) = 2MPDA = 22.00 - 19.00 = \;\;3.00$

$Z = 20.50$

$MPDA = 1.50$

Adding the two estimates together cancels out the effect of *MPDA* and dividing this result by 2 yields an unconfounded estimate of *Z*. Similarly subtracting the two estimates cancels out the *Z* effect leaving *MPDA* unconfounded. This example illustrates that it is possible to use fractional factorial designs in blocks. This is especially important for smaller fractions and this point will be picked up later in Section 4.3.3.

As a general rule, when large numbers of runs are required in a full factorial design (*f* is large) and these have to carried out sequentially, it is almost always better to run a half-fraction containing the first sixteen runs, analyse the results, consider the implications and then, if necessary, carry out the second half. Often, however, it might be more valuable to introduce new factors or change the levels of the factors. Such a sequential approach can accelerate progress. Note that the runs should be randomized within the fractions each time an experiment is conducted. If both half-fractions are carried out the design is then a randomized blocked factorial design. Some information concerning the interaction confounded with the block will therefore have been lost.

4.3.2. Generating Quarter-fraction Factorial Designs

Suppose now you have not carried out the half-fraction factorial design above but instead have investigated the effects of six factors in the same number of experimental runs (16), the previous five factors plus an additional factor, temperature (*T*). The design now constitutes a 2^{6-2} design, that is a quarter replicate of the full factorial design for six variables. Obviously, this design represents a great saving on the 64 runs required for the full factorial design. The original five factors retain the aliases worked out above. However, each effect now has additional aliases, the identities of which can be found with additional defining relations. For this quarter-fraction we need to define the levels of the sixth factor, temperature (*T*). We can do this by letting it equal *ZMP* so that where *ZMP* has a minus sign then *T* is set at its lower level and so on. We now have two generators

$$A = ZMPD \qquad \text{and} \qquad T = ZMP$$

The defining relations associated with these are $I = ZMPDA$ and $I = ZMPT$ respectively.

A third defining relation is now obtained by multiplying these defining relations together and cancelling out the elements which appear an even number of times.

$$I = ZMPDA \times ZMPT = Z^2M^2P^2DAT = DAT$$

∏   Calculate the aliases for the quarter replicate and decide which of the following are not aliases of ZM? (PDA, PAT, $ZMDAT$, $ZMPT$, PT)

The defining relations given above should have helped you see that PDA, $ZMDAT$ and PT are aliases whereas PAT and $ZMPT$ are not.

$$I = ZMPDA, I = ZMPT, I = DAT$$

Thus

$$ZM = ZM \times ZMPDA = PDA$$
$$ZM = ZM \times ZMPT = PT$$
$$ZM = ZM \times DAT = ZMDAT$$

If you managed to work out the aliases correctly you will have seen that ZM only has three aliases (the design being a quarter replicate). This is true of all the effects which are given in full below:

Aliases							Estimate
Average	=	ZMPDA	=	ZMPT	=	DAT	116.00
Z	=	MPDA	=	MPT	=	ZDAT	22.00
M	=	ZPDA	=	ZPT	=	MDAT	−0.50
P	=	ZMDA	=	ZMT	=	PDAT	1.50
D	=	ZMPA	=	ZMPDT	=	AT	10.75
A	=	ZMPD	=	ZMPAT	=	DT	−7.75
ZM	=	PDA	=	PT	=	ZMDAT	3.00
ZP	=	MDA	=	MT	=	ZPDAT	3.00
ZD	=	MPA	=	MPDT	=	ZAT	9.25
ZA	=	MPD	=	MPAT	=	ZDT	−0.25
MP	=	ZDA	=	ZT	=	MPDAT	2.00
MD	=	ZPA	=	ZPDT	=	MAT	−2.25
MA	=	ZPD	=	ZPAT	=	MDT	−0.25
PD	=	ZMA	=	ZMDT	=	PAT	−1.25
PA	=	ZMD	=	ZMAT	=	PDT	0.75
DA	=	ZMP	=	ZMPDAT	=	T	−8.00

4.3.3. Separating the Aliases

In the half-fraction design each main effect was aliased with a four-factor effect. However, in the quarter fraction some main effects are aliased with two-factor effects. It is likely that two-factor effects will be more important than four-factor effects and there is, therefore, an increased risk that where an estimate of a main effect is large it may not be due entirely to the factor but instead to one of its aliases. Therefore, we have to make sure when we are designing these experiments that main effects are confounded with the highest possible order of effects.

In the above 2^{6-2} quarter-fraction design each estimate was shown to represent a group of four effects. Two quarter-fraction designs can together form a half-factorial design in which each effect is only confounded with one other effect. Thus adding a second fraction produces a degree of separation of the effects. There is usually a choice of design for this second set which I will now endeavour to show.

The defining relations used to produce the estimates for the quarter-fraction were

$$I = ZMPDA = ZMPT = DAT$$

generated by letting $A = ZMPD$ and $T = ZMP$. The result was that some of the main effects were confounded with two-factor effects, e.g.

$$A + ZMPD + ZMPAT + DT$$

Suppose now you wish to carry out another quarter-fraction which will enable you to separate the main effects from two-factor interactions. What defining relations should you use? Well, the first quarter-fraction had two defining relations which multiplied together to produce the third. Of course it is possible to use negative signs in either of the generators to produce these defining relations. In total four sets of defining relations are possible (four quarter-fractions):

$$
\begin{array}{llll}
I = & ZMPDA = & ZMPT = & DAT & \text{(i)} \\
I = & -ZMPDA = & -ZMPT = & DAT & \text{(ii)} \\
I = & ZMPDA = & -ZMPT = & -DAT & \text{(iii)} \\
I = & -ZMPDA = & ZMPT = & -DAT & \text{(iv)}
\end{array}
$$

The first of these is the original one, but which one should we now use to de-alias the main effects from the two-factor interactions? Let us go back to the estimates of the effects of factor A and its aliases. These are given below for the four possible quarter-fractions:

$$A + ZMPD + ZMPAT + DT = (x_1)$$
$$A - ZMPD - ZMPAT + DT = (x_2)$$
$$A + ZMPD - ZMPAT - DT = (x_3)$$
$$A - ZMPD + ZMPAT - DT = (x_4)$$

Suppose we carry out fraction (iii) and end up with an estimate for A and its aliases given by (x_3). We could add (x_1) and (x_3) to give:

$$A + ZMPD + ZMPAT + DT = (x_1)$$
$$A + ZMPD - ZMPAT - DT = (x_3)$$
$$\overline{\qquad\qquad\qquad\qquad\qquad\qquad}$$
$$2 \times (A + ZMPD) \qquad = (x_1 + x_3)$$

Averaging the estimates $(x_1 + x_3)$ yields an estimate for A which is free of two-factor interactions.

∏   What other fraction would also yield an estimate for A which is also free of two-factor interactions?

A quick glance at the signs in front of the estimates should have told you that using the second defining relation (ii) would be of no use to you because both fractions (i) and (ii) have + signs in front of DAT. This is the alias responsible for A being confounded with a two-factor interaction (DT). This leaves fraction (iv) as the only other possibility. You can confirm this for yourself by repeating the above procedure. Alternatively, you might have spotted that fraction (iv) has a − sign for DAT which would cancel when added to (i).

So, we now have two fractions which leave A de-aliased from DT. Let us now look at another main effect (D) which, given the above four defining relations, would have the following aliases:

$$D + ZMPA + ZMPDT + AT = (x_1)$$
$$D - ZMPA - ZMPDT + AT = (x_2)$$
$$D + ZMPA - ZMPDT - AT = (x_3)$$
$$D - ZMPA + ZMPDT - AT = (x_4)$$

`Averaging the responses from fractions (i) and (iii) or (i) and (iv) will again leave the main effect de-aliased from the two-factor interactions by cancelling the *AT* effect. This is true of all the main effects. Therefore, either fraction (iii) or (iv) will suffice as an additional quarter-fraction which will de- alias the main effects from two-factor interactions.

SAQ 4.3b	Suppose we decided to investigate the following seven factors at two levels in eight experimental runs in a reactor:

Factor	–	+
1 Temperature (°C)	120	160
2 Concentration (%)	5	10
3 Catalyst (%)	0.5	1.0
4 Slug (Y/N)	N	Y
5 Stirring rate (rpm)	100	150
6 Feed rate (l/min)	10	12
7 Catalyst (A/B)	A	B

The most appropriate design for such a situation would seem to be a 2^{7-4} design (a one-sixteenth fraction). There are sixteen generators available to us. I have selected the following generators:

$$4 = 12, 5 = 13, 6 = 23, 7 = 123$$

which gives the defining relations

$$I = 124 = 135 = 236 = 1237$$

A full set of defining relations can only be worked out by multiplying all the possible combinations of defining relations. Therefore, multiplying two of these at a time gives $\longrightarrow$

SAQ 4.3b
(cont.)

$I = 2345 = 1346 = 347 = 1256 = 257 = 167$

and multiplying three at a time gives

$I = 456 = 1457 = 2467 = 3567$

and four at a time gives

$I = 1234567$

The complete set of defining relations is therefore

$I = 124 = 135 = 236 = 1237 = 2345 = 1346 = 347$
$= 1256 = 257 = 167 = 456 = 1457 = 2467 = 3567$
$= 1234567$

The estimates of the main effects would have several two-factor effects aliased to them such that:

$1 + 24 + 35 + 67 = (E_1)$
$2 + 14 + 36 + 57 = (E_2)$
$3 + 15 + 26 + 47 = (E_3)$
$4 + 12 + 56 + 37 = (E_4)$
$5 + 13 + 46 + 27 = (E_5)$
$6 + 23 + 45 + 17 = (E_6)$
$7 + 34 + 25 + 16 = (E_7)$

Work out another two sets of generators for one-sixteenth fractional replicates which will

(i) de-alias all the main effects from all two-factor interactions.

(ii) de-alias factor *4* from all two-factor interactions.

4.4. DESIGN RESOLUTION

4.4.1. Best Fractions

The original 2^{5-1} half-fraction example had a confounding pattern in which the main effects were confounded by four-factor interactions and two-factor interactions with three-factor interactions. This design is a *Resolution* V (five) design because the defining relation has five letters. If there is more than one defining relation then it is the length of the shortest defining relation which actually controls the resolution of the design. It is possible to set up designs so that main effects are confounded by interactive effects of varying order. For example, if a half-fraction had been generated by letting $A = ZMP$ then some main effects would be confounded by three-factor interactions and the design would now be of resolution IV. Since it is likely that three-factor interactions are more important than four-factor interactions this design is not as good as the original design which is termed the *best* half-fraction. In general a design of resolution R is one in which no p-factor is confounded with any effect containing less than $R-p$ factors. The resolution of a design is denoted by the appropriate roman numeral either as a subscript or after the fraction. For example, the two 2^{5-1} half-fractions given above are either

$$2^{5-1}(V)$$

or

$$2^{5-1}(IV)$$

∏   What is the resolution of a half-fraction of a four-factor fractional
      factorial design (factors *1*, *2*, *3* and *4*) which has *4 = 123* as its
      generator? Is it possible to produce a better design?

The quickest way to sort out the resolution of this design is by making the
defining relation, which must be $I = 1234$. This is the only defining relation
and therefore contains the shortest word. The design is then designated as

$$2^{4-1}(IV)$$

This is in fact the best fraction that can be generated for half-fractions
of a 2^4 design. If the generator had been *4 = 12* or any other two-
factor interaction, the main effects would have been aliased with two-factor
interactions and two-factor interactions aliased with main effects instead of
some main effects having three-factor effects as their aliases and two-factor
interactions being aliased with each other.

There are some general rules governing resolution which are given below:

1. A design of resolution $R = $ III does not confound main effects with one
   another, but does confound main effects with two-factor interactions.

2. A design of resolution $R = $ IV does not confound main effects with
   other main effects or two-factor interactions but does confound two-
   factor interactions with other two-factor interactions.

3. A design of resolution $R = $ V does not confound main effects and
   two-factor interactions with each other, but does confound main effects
   with four-factor interactions and two-factor interactions with three-factor
   interactions and so on.

To obtain any half-fraction of the highest resolution the same procedure
as used for the 2^{5-1} half-fraction is followed. Simply write a full factorial
design for the first $f - 1$ factors and associate the fth factor with plus or
minus the highest-order interaction column.

All the fractional designs considered up to now have been based on two-level factorials to assess the effects of any number of factors in N runs, where N is a power of 2. These enable us to have saturated designs of resolution III. For example, the 2^{7-4} design can estimate seven main effects in eight runs but is of resolution III and has its main effects confounded with two-factor interactions. This is a saturated design because it is not possible to increase the level of confounding. These saturated designs are most effective when none of the interactions are important, leaving the main effect estimates unbiased. Plackett and Burmann developed saturated designs of resolution III in which N is a multiple of 4. An example of this is a design in which 11 factors can be examined in 12 runs.

SAQ 4.4a   What is the resolution of the quarter-fraction generated by introducing the sixth factor (Temperature) into the 2^{5-1} half-fraction so as to give the 2^{6-2} design with the defining relations $I = ZMPDA$, $I = ZMPT$ and $I = DAT$? Generate a "better" quarter-fraction design.

4.4.2. Complete Factorials Embedded in Fractions

The 2^{5-1} (V) half-fraction with the relation $I = ZMPDA$ was obtained by associating A with $ZMPD$. If the main effect of A and all its interactions turned out to be unimportant we would effectively have a complete factorial design for factors $ZMPD$ in 16 runs. This is the same as if all the columns containing A in the fractional design matrix had been omitted. If, however, you were to try omitting one of the other factors, such as pH (P), then this too reduces the design to a complete factorial in the remaining factors ($ZMDA$). The original justification for using a fraction of the runs was that the higher-order interactions were likely to be unimportant. We now have an alternative justification that probably not all the factors are going to influence the response and that complete factorials in the remaining factors will be generated.

In Part 3 cube diagrams of factorial designs were introduced to show how main effects and interactions are estimated. They also serve a purpose in showing the geometry of fractional factorial designs. If a 2^{3-1} fractional design is set up, four runs are required. This design is shown in Fig. 4.4a.

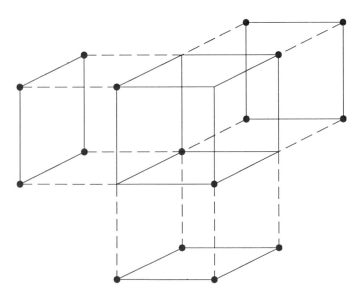

Fig. 4.4a. *A 2^{3-1} (III) fractional factorial design showing projections in 2^2 factorial designs when factors are unimportant*

If one of the factors does not influence the response the half-fraction design collapses from three dimensions into a complete 2^2 factorial design in the remaining factors.

To illustrate this, consider the 2^{5-1} half-fraction carried out for the five factors $(Z, M, P, D$ and $A)$ in which two main effects and three interactions were found to be important. These effects involved only three factors in total $(Z, D$ and $A)$. The design can now be visualized in cube diagrams as a replicated 2^3 design (Fig. 4.4b) and the main effects of these factors estimated by contrasting faces on opposite sides of the cube, as shown in Part 3.

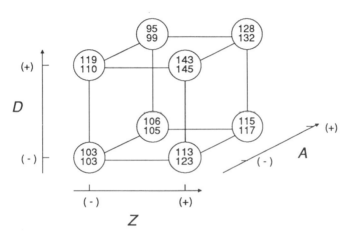

Fig. 4.4b. *Responses from the 2^{5-1} half-fraction factorial design, phosphatase example shown as a 2^3 factorial design in factors Z, D and A with replicated responses*

SAQ 4.4b

Complete the following paragraph by inserting the most appropriate word or phrase, chosen from the list below, into the blank spaces.

When the number of .... to be investigated in an experiment is large, the number of experimental runs required is often prohibitive and many of the ...... are likely to be .... Fractional factorial designs are a useful way of exploiting this, especially when little is known about the system being investigated. A "best" half-fraction 2^{5-1} design in factors *1, 2, 3, 4* and *5* can be .... by first writing down a full design for ...... factors and letting the signs indicating the levels of the fifth factor equal the signs for the four-factor interaction. Such a half-fraction design requires 16 runs. However, each estimate is now a combination of ...... effects, which are .... These effects are said to be aliases of one another. In this 2^{5-1} design the ...... effects are confounded by four-factor effects, and the ......-factor effects are confounded by two-factor effects. A 2^{5-1} half-fraction with the same .... could have been produced by the generator *5 = −1234*.

FACTORS	DEFINING RELATION
ALIASES	RESOLUTION
EFFECTS	TWO
THREE	FOUR
CONFOUNDED	FIVE
REDUNDANT	GENERATED
MAIN	

SAQ 4.4c

Consider a 2^{7-3} fractional factorial design.

(a) How many variables does this design have?

(b) How many runs does it require?

(c) How many defining relations are there?

(d) How many independent generators are there for the design?

(e) What is the highest possible resolution of the design?

4.5.   ESTIMATION, ANALYSIS AND INTERPRETATION OF EFFECTS

The effects in a fractional factorial experiment can be estimated in exactly the same way as factorial designs including from first principles, columns of contrast coefficients or from Yates' algorithm. If I am doing this without the aid of computer software I personally prefer to use Yates' algorithm, since this can give the sum of squares values, which may be used later in an *ANOVA*.

As a general rule, when you wish to investigate the effects of a large number of factors in fractional designs using few experimental runs you should try to select a design which gives you the highest resolution, that is, one in which the main effects are confounded with the higher-order interactions. However, these designs are really only applicable if you can guess or know from theoretical considerations that not all the factors are likely to be important, so that the designs collapse into fewer dimensions. Also, if you think certain interactions are likely to be important before the experimental runs have been carried out you should select a design which keeps this interaction free of all main effects.

It is often possible to adjust the ranges of the factors which you are examining. As a general rule interactions become small compared to main effects when the ranges of factors are decreased, but experimental error is likely to become relatively more important. Therefore, it is best to select ranges which produce effects which you can measure, but are not so large as to give appreciable interactions. Careful examination of past experimental records or performance of your experimental system may provide you with this information.

When using fractional factorial designs most of the available degrees of freedom are used up in estimating main effects and two-factor interactions, leaving very few to estimate the residual error variance. As with unreplicated factorial designs, which have no external estimate of residual error variance, you need to use information from within the experiment to provide you with this estimate. If main effects are confounded with two-factor interactions and an analysis of variance is required, you have to assume that at least one of the main effects is going to be negligible (probably quite a dangerous assumption). Also the fewer the number of degrees of freedom that are used in the residual error the less sensitive will the subsequent analysis be.

Fractional factorial designs are not usually an end to experimentation. You can then run them in sequence. If you select generators carefully the information from the experiments can be combined, leaving you, for example, with de-aliased main effects. The rest of the estimates may then be combined to form the residual error variance for use in an *ANOVA*.

Learning Objectives

After studying the material in Part 4 you should be able to

- understand the limitation in the number of factors that can be successfully examined in a full factorial design;

- generate half-fraction designs and work out the aliases;

- work out the defining relations for higher-fraction designs;

- decide which fractions to run when de-aliasing effects;

- understand the concept of design resolution.

SAQs AND RESPONSES FOR PART FOUR

SAQ 4.2   What fraction of a complete two-level factorial design is necessary to investigate the effects of eight factors in 16 runs?

A: 1/2   B: 1/10

C: 1/8   D: 1/16

Response

If you chose A, you chose incorrectly because a half-fraction of a 2^8 factorial design requires the same number of runs as a 2^7 complete factorial design, that is 128 runs. If you considered B as a possible answer for more than about a quarter of a second you'd better go back to the beginning of Part 4!! Answer C is also incorrect because

$$1/8 \times 2^8 = 2^8 - 2^3 = 2^5 = 32 \text{ runs}$$

whereas you wanted an experiment with only 16 runs. The correct answer (*D*) can be obtained most easily if you work out what fraction 16 is of 2^8, that is

$$16/256 = 1/16$$

****

SAQ 4.3a	A combination of these two half-fractions yields the complete factorial design for five variables at two levels [True/False]?

Response

I trust you stated "true" for this answer. This is most easily seen if the two half-fractions are written out in full.

	A = ZMPD								A = −ZMPD						
	Factor								Factor						
Run	Z	M	P	D	A	ZMPD	Response	Run	Z	M	P	D	A	ZMPD	Response
17	−	−	−	−	+	+	106	1	−	−	−	−	−	+	109
2	+	−	−	−	−	−	113	18	+	−	−	−	+	−	120
3	−	+	−	−	−	−	103	19	−	+	−	−	+	−	113
20	+	+	−	−	+	+	115	4	+	+	−	−	−	+	113
5	−	−	+	−	−	−	103	21	−	−	+	−	+	−	109
22	+	−	+	−	+	+	117	6	+	−	+	−	−	+	104
23	−	+	+	−	+	+	105	7	−	+	+	−	−	+	106
8	+	+	+	−	−	−	123	24	+	+	+	−	+	−	115
9	−	−	−	+	−	−	119	25	−	−	−	+	+	−	96
26	+	−	−	+	+	+	128	10	+	−	−	+	−	+	146
27	−	+	−	+	+	+	95	11	−	+	−	+	−	+	111
12	+	+	−	+	−	−	143	28	+	+	−	+	+	−	127
29	−	−	+	+	+	+	99	13	−	−	+	+	−	+	116
14	+	−	+	+	−	−	145	30	+	−	+	+	+	−	131
15	−	+	+	+	−	−	110	31	−	+	+	+	+	−	92
32	+	+	+	+	+	+	132	16	+	+	+	+	−	+	148

Note that the designs are the same for Z, M, P, D and $ZMPD$. However, in the first half-fraction, A has the same signs as $ZMPD$ so that these estimates will be confounded. In the second half-fraction the signs for A and $ZMPD$ are opposite so that the alias of A is $-ZMPD$. Combining the two design matrices therefore gives columns with different signs for both estimates. This applies to all the aliases.

****

SAQ 4.3b

Suppose we decided to investigate the following seven factors at two levels in eight experimental runs in a reactor:

Factor	–	+
1 Temperature (°C)	120	160
2 Concentration (%)	5	10
3 Catalyst (%)	0.5	1.0
4 Slug (Y/N)	N	Y
5 Stirring rate (rpm)	100	150
6 Feed rate (l/min)	10	12
7 Catalyst (A/B)	A	B

The most appropriate design for such a situation would seem to be a 2^{7-4} design (a one-sixteenth fraction). There are sixteen generators available to us. I have selected the following generators:

$$4 = 12, 5 = 13, 6 = 23, 7 = 123$$

which gives the defining relations

$$I = 124 = 135 = 236 = 1237$$

A full set of defining relations can only be worked out by multiplying all the possible combinations of defining relations. Therefore, multiplying two of these at a time gives

$$I = 2345 = 1346 = 347 = 1256 = 257 = 167$$

and multiplying three at a time gives

$$I = 456 = 1457 = 2467 = 3567$$

$\longrightarrow$

SAQ 4.3b
(cont.)

and four at a time gives

$$I = 1234567$$

The complete set of defining relations is therefore

$$I = 124 = 135 = 236 = 1237 = 2345 = 1346 = 347$$
$$= 1256 = 257 = 167 = 456 = 1457 = 2467 = 3567$$
$$= 1234567$$

The estimates of the main effects would have several two-factor effects aliased to them such that:

$$1 + 24 + 35 + 67 = (E_1)$$
$$2 + 14 + 36 + 57 = (E_2)$$
$$3 + 15 + 26 + 47 = (E_3)$$
$$4 + 12 + 56 + 37 = (E_4)$$
$$5 + 13 + 46 + 27 = (E_5)$$
$$6 + 23 + 45 + 17 = (E_6)$$
$$7 + 34 + 25 + 16 = (E_7)$$

Work out another two sets of generators for one-sixteenth fractional replicates which will

(i) de-alias all the main effects from all two-factor interactions.

(ii) de-alias factor 4 from all two-factor interactions.

Response

This design constitutes a *saturated design* because all the main effects are confounded with two-factor interactions. The generators used for the original fraction all have plus signs;

$$4 = 12, 5 = 13, 6 = 23, 7 = 123$$

which yield the defining relations given previously. A full set of main effects and two-factor aliases has been worked out and we first need to consider how to de-alias all these main effects from the two-factor interactions, that is to find out which of the remaining fifteen generators should be selected. If you understood this section it should be fairly clear to you that changing the signs of all the generators in which main effects are confounded with two-factor interactions will alter the signs of all the aliased two-factor effects. Thus,

$$4 = -12, 5 = -13, 6 = -23, 7 = 123$$

which gives the defining relations

$$I = -124 = -135 = -236 = 1237$$

A complete set of defining relations for this is obtained by multiplying these defining relations together two at a time to give:

$$I = 2345 = 1346 = -347 = 1256 = -257 = -167.$$

Multiplying three at a time gives

$$I = -456 = 1457 = 2467 = 3567.$$

And four at a time gives

$$I = -1234567.$$

A partial set of defining relations for the main effects confounded with two-factor effects is therefore

$$I = -124 = -135 = -236 = -347 = -257 = -167 = -456.$$

This will alter the signs of all the two-factor interactions so that the estimates now have the following aliases:

$$1 - 24 - 35 - 67 = (E_1')$$
$$2 - 14 - 36 - 57 = (E_2')$$
$$3 - 15 - 26 - 47 = (E_3')$$
$$4 - 12 - 56 - 37 = (E_4')$$
$$5 - 13 - 46 - 27 = (E_5')$$
$$6 - 23 - 45 - 17 = (E_6')$$
$$7 - 34 - 25 - 16 = (E_7')$$

Obviously running a second fractional experiment with this defining relation will enable us to add these new estimates (in which all the two-factor effects are negative) to the original estimates (in which they are all positive). Dividing the results by 2 will yield the unconfounded main effect estimates.

For example, adding (E_1) to (E_1') will give the following:

$$1 + 24 + 35 + 67 \quad = (E_1)$$
$$1 - 24 - 35 - 67 = (E_1')$$
$$\overline{}$$
$$1 \qquad\qquad\qquad = 0.5 \times (E_1 + E_1')$$

Clearly the same will be true of all the other estimates. Adding a new fraction in which the signs of all the generators have been reversed will leave the main effect estimates free of two-factor interactions. This is a fairly powerful tool, especially when fractions can be carried out in sequence. It is possible to direct the fractions to de-alias some of the effects in which you are interested.

In the same way as changing all the signs of the generators de-aliased all main effects from two-factor interactions, it is possible to select particular effects which you wish to de-alias. This is the second problem you have been set.

To solve this you again have to take the original generators:

$$4 = 12, 5 = 13, 6 = 23, 7 = 123$$

in which for this defining relation factor *4* is aliased with *12*. If you change this sign it obviously changes the sign of all the defining relations involving factor *4*. Thus

$$I = -124 = 135 = 236 = 1237.$$

Multiplying two at a time together gives

$$I = -2345 = -1346 = -347 = 1256 = 257 = 167.$$

Multiplying three at a time gives

$$I = -456 = -1457 = -2467 = 3567.$$

And four at a time gives

$$I = -1234567.$$

A partial defining relation for the main effect of factor 4 confounded with two-factor effects is therefore

$$I = -124 = 135 = 236 = -347 = 257 = 167 = -456.$$

Working out the aliases of all the estimates now reveals what has happened;

$$
\begin{aligned}
1 - 24 + 35 + 67 &= (E_1') \\
2 - 14 + 36 + 57 &= (E_2') \\
3 + 15 + 26 - 47 &= (E_3') \\
4 - 12 - 56 - 37 &= (E_4') \\
5 + 13 - 46 + 27 &= (E_5') \\
6 + 23 - 45 + 17 &= (E_6') \\
7 - 34 + 25 + 16 &= (E_7')
\end{aligned}
$$

Any effect involving factor 4 now has a − sign. If this fraction is run and the estimates $E_4 + E_4'$ averaged, an unconfounded estimate of the effect of factor 4 will be obtained.

****

SAQ 4.4a	What is the resolution of the quarter-fraction generated by introducing the sixth factor (Temperature) into the 2^{5-1} half-fraction so as to give the 2^{6-2} design with the defining relations $I = ZMPDA$, $I = ZMPT$ and $I = DAT$? Generate a "better" quarter-fraction design.

Response

If you thought it was a resolution V design then you were incorrect because in such a design all the main effects are confounded by four-factor and higher-order effects. However, the third defining relation given above ($I = DAT$) shows that D is confounded by the interaction AT. You are also incorrect if you thought that it is resolution IV because the lowest-order effects by which main effects are confounded now are three-factor interactions. The correct answer is resolution III simply because the shortest defining relation is three letters in length ($I = DAT$).

If possible, a better quarter-fraction design would have a higher resolution (IV or possibly V). To generate the original 2^{5-1} half-fraction it was apparently logical to set the fifth factor (A) to be confounded by the four-factor interaction $ZMPD$. However, introduction of the sixth factor (T), as $T = ZMP$, led to the formation of the defining relation $I = DAT$ and the quarter-fraction design was consequently resolution III. One way to avoid such an occurrence is to select the generators carefully. If the generators had been $A = ZMP$ and $T = MPD$, the two set defining relations would then have been $I = ZMPA$ and $I = MPDT$, which would then multiply together to form the third relation $I = ZDAT$. The shortest word would now be four letters and the design one of resolution IV. This shows that careful selection of generators can lead to improvements in the resolving power of fractional designs. This should influence your thinking when considering if you are going to enhance the number of factors you intend to examine in 16 runs. Don't unnecessarily generate the best half-fraction and add in extra factors.

$$**********************************$$

SAQ 4.4b

Complete the following paragraph by inserting the most appropriate word or phrase, chosen from the list below, into the blank spaces.

When the number of .... to be investigated in an experiment is large, the number of experimental runs required is often prohibitive and many of the ...... are likely to be .... Fractional factorial designs are a useful way of exploiting this, especially when little is known about the system being investigated. A "best" half-fraction 2^{5-1} design in factors *1*, *2*, *3*, *4* and *5* can be .... by first writing down a full design for ...... factors and letting the signs indicating the levels of the fifth factor equal the signs for the four-factor interaction. Such a half-fraction design requires 16 runs. However, each estimate is now a combination of ...... effects, which are .... These effects are said to be aliases of one another. In this 2^{5-1} design the ...... effects are confounded by four-factor effects, and the ......-factor effects are confounded by two-factor effects. A 2^{5-1} half-fraction with the same .... could have been produced by the generator *5 = −1234*.

FACTORS	DEFINING RELATION
ALIASES	RESOLUTION
EFFECTS	TWO
THREE	FOUR
CONFOUNDED	FIVE
REDUNDANT	GENERATED
MAIN	

Response

When the number of FACTORS to be investigated in an experiment is large, the number of experimental runs required is often prohibitive and many of the EFFECTS are likely to be REDUNDANT. Fractional factorial designs are a useful way of exploiting this, especially when little is known about the system being investigated. A "best" half-fraction 2^{5-1} design in factors *1*, *2*, *3*, *4* and *5* can be GENERATED by first writing down a full design for FOUR factors and letting the signs indicating the levels of the fifth factor equal the signs for the four-factor interaction. Such a half-fraction design requires 16 runs. However, each estimate is now a combination of TWO effects, which are CONFOUNDED. These effects are said to be aliases of one another. In this 2^{5-1} design the MAIN effects are confounded by four-factor effects and the THREE-factor effects are confounded by two-factor effects. A 2^{5-1} half-fraction with the same RESOLUTION could have been produced by the generator $5 = -1234$.

$$*******************************$$

SAQ 4.4c	Consider a 2^{7-3} fractional factorial design.
	(a) How many variables does this design have?
	(b) How many runs does it require?
	(c) How many defining relations are there?
	(d) How many independent generators are there for the design?
	(e) What is the highest possible resolution of the design?

Response

The number of factors investigated here is seven, as given by the first superscript after the base number. The number of runs required is 16, which is the number of levels (2) to the power of $7-3 = 4$. The number of runs is thus 2^4 or 16. To obtain the number of words in the defining relation, think of the fraction that the design is. The design investigates the effects of seven factors in 16 runs. A complete 2^7 factorial design normally requires 128 runs. The 2^{7-3} design is therefore an eighth fraction (16/128) in which each effect has seven aliases. Thus, including I, there are eight defining relations. The number of independent generators is found by designing a complete factorial for the number of runs available and then letting subsequent generators equal various interactions up to the number of factors to be investigated. For this one-eighth fraction the first four factors are set as for a complete factorial design, and the fifth, sixth and seventh factors from interactions. Therefore, a 2^{7-3} design requires three independent generators. Designating the factors as 1 to 7 the "best-fraction" available is generated as follows;

$$5 = \pm \; 123$$
$$6 = \pm \; 234$$
$$7 = \pm \; 134$$

The defining relations associated with these generators are then

$$I = 1235$$
$$I = 2346$$
$$I = 1347$$

Further relations produced from these are:

$$I = 1235 \times 2346 \qquad = 1456$$
$$I = 1235 \times 1347 \qquad = 2457$$
$$I = 2346 \times 1347 \qquad = 1267$$
$$I = 1235 \times 2346 \times 1347 = 3567$$

Each of these generators contains at least four numbers, making the design resolution IV.

$$*****************************$$

5. Response Surface Methodology

In this part of the book I am going to introduce you to the types of experimental design which can be used for investigating response surfaces and optimizing responses. There are many possible applications of these techniques in chemistry. They may be used, for example, to optimize the yield of a reaction, to decrease the level of impurity present in an end product or to improve a chromatographic separation of solutes. In the first few sections I will also introduce some of the matrix algebra which is necessary for you to be able to use these techniques. It is not intended as a comprehensive insight into matrix algebra, but should enable you to carry out some of the basic manipulations and to understand other more complex manipulations.

5.1. RESPONSE SURFACES

First of all it might be useful to define for you what I mean by a response surface. One fairly general definition is that a response surface is a graph of a response as a function of one or more factors. Whatever this response may be, most response surface methods work best when the response and the factors vary in a continuous manner, that is, they may take any value within a specific range. For example, a response surface design may be applied to find the optimum combination of temperature and amount of a reactant which maximizes the percentage yield of a product.

It is possible to optimize more than one response parameter at a time by developing what are termed composite response functions. However, consideration of these at this point would probably deflect you from some of the more important aspects of response surface investigation. For this reason only one response parameter will be considered in this treatment of the subject.

If one factor (x) is varied in a particular experimental design, the response (y) may be plotted against the level of the factor in the form of a simple two dimensional graph, with the response on the y axis and the factor levels on the x axis. If the factor is examined at two levels it may only be modelled as a straight line (Fig. 5.1a).

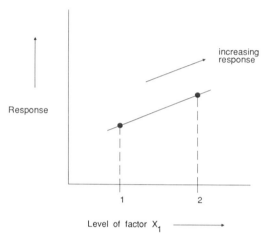

Fig. 5.1a. *Response surface for one factor at two levels*

However, if the design allows three or more levels of a factor to be selected, curvature in the response may also be estimated (Fig. 5.1b). Of course there might be a large number of possible values that factor x could take, but in practice only a few of these are used, giving an incomplete picture of the response surface. Most commonly it is assumed that a functional relationship of some sort exists between the response and the factor. The experiment is carried out and a model is fitted to the data. The model may be mechanistic, such as with the Beer-Lambert spectrometric law ($A = \epsilon lc$), in which every aspect of the relationship is understood. Alternatively, it may be empirical in nature, in which the relationship between the response and the factor(s) is poorly understood and has to be modelled by developing some general rules.

The techniques I will be introducing in this part are used under the assumption that little is known about the functional mechanism between

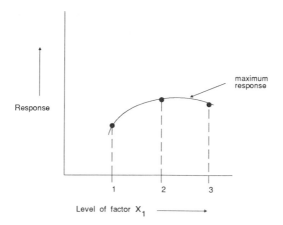

Fig. 5.1b. *Response surface for one factor at three levels*

the response and the factor level. Indeed for a relationship to be understood clearly, if at all possible, an inordinate number of experimental runs may have to be conducted, at the expense of valuable time and effort. In contrast, the response surface methods that I will introduce are fairly efficient and may give you a reasonable idea about the response surface and a likely position of an optimum in very few experimental runs.

5.2. SINGLE-FACTOR FIRST-ORDER MODEL

When assuming a functional relationship between a response and a factor the subsequent model has parameters that should fit the data reasonably well. The task is to estimate these parameters. Suppose the situation is as in Fig. 5.1a and only two levels of the factor have been investigated. Only the straight line slope and intercept may be calculated. The appropriate model is first-order:

$$y_i = \beta_0 + \beta_1 x_{1i} \qquad (5.1)$$

where y_i is the response for the ith run, β_0 is the y intercept at factor level $x_1 = 0$, β_1 is the straight line slope for factor x_1, and x_{1i} is the level taken by factor x_1 in the ith run. You should have noticed this is the same as the common expression

$$y = mx + c.$$

I will use the β notation here because it is in common use and also allows the model to be expanded to include other factors, curvature and interactions. However, this model assumes that any response is going to fall exactly on that line (deterministic model), a highly unlikely event in any experimental work. More often the line does not perfectly fit the data. Vertical differences between the data points and the fitted line are the residuals or residual errors as shown in Fig. 5.2.

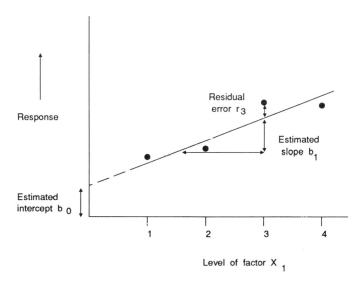

Fig. 5.2. *Possible fit of hypothetical data for the first-order model* $y_i = \beta_0 + \beta_1 x_{1i} + r_i$ *demonstrating additivity of terms*

The model now has to change to incorporate the possibility of error in the response and becomes a probabilistic or statistical model;

$$y_i \; = \; \beta_0 + \beta_1 x_{1i} + r_i \tag{5.2}$$

where the terms are as previously defined with the addition that r_i is the residual error for the ith experimental run. The model is in fact a linear model, not because I fitted a straight line to the data, but because it is made up of additive terms which have at most one multiplicative term in them. The observed response is considered to be the combination of some response due to the intercept β_0, some response due to the run being carried out at a particular level of factor x_1, plus some residual error (Fig. 5.2).

This definition of linear models allows us to expand the models to include curvature, effects due to other factors, and also interactions, yet still remain a linear equation. Thus

$$y_i = \beta_0 + \beta_1 x_{1i} + \beta_2 x_{2i} + \beta_{12} x_{1i} x_{2i} + r_i$$

is a linear model which includes an effect due to a second factor x_2 and a second-order interaction term $(\beta_{12} x_{1i} x_{2i})$.

∏   Which of the following models are linear?

(a) $y_i = \beta_0 + r_i$

(b) $y_i = \beta_1 \beta_2 x_{1i} x_{2i} + r_i$

(c) $y_i = \beta_0 + \beta_1 x_{1i} + \beta_{11} x_{1i}^2 + \beta_{111} x_{1i}^3 + r_i$

(d) $y_i = \beta_0 10^{\beta_1 x_{1i}}$

Models (a) and (c) are linear because they only contain one β in every term. Models (b) and (d), on the other hand, have terms containing more than one β. Model (d) could be transformed to a linear model by taking the $\log_{10}$ of both sides of the equation to give

$$\log y_i = \log \beta_0 + \beta_1 x_{1i}$$

5.2.1. Single-factor First-order Example

In this introductory section on response surfaces an example of a single-factor first-order model will be introduced to illustrate some of the more relevant aspects of response surface techniques. Matrix algebra will be used to calculate the required quantities.

Suppose that a chemist was interested in investigating the effect of reactant concentration upon reaction yield in which four levels of reactant concentration (1%, 2%, 3% and 4%) were to be used. The order of carrying out the experimental runs was randomized and the product yield determined as a percentage of the theoretical maximum. These are given below:

Yield (%)	Reactant Concentration (%)
10	1
15	2
20	3
23	4

It is possible to define a design matrix D for this experiment which contains the levels specified in each of the runs such that

$$D = \begin{bmatrix} 1 \\ 2 \\ 3 \\ 4 \end{bmatrix}$$

A first-order model of the type given in Eq. (5.2) was applied so that each response (y_i) can be represented as a linear sum of effects due to an intercept β_0, a slope β_1, and a residual error r_i.

$$10 = 1 \times \beta_0 + 1 \times \beta_1 + r_1$$
$$15 = 1 \times \beta_0 + 2 \times \beta_1 + r_2$$
$$20 = 1 \times \beta_0 + 3 \times \beta_1 + r_3$$
$$23 = 1 \times \beta_0 + 4 \times \beta_1 + r_4$$

These equations can be viewed as four simultaneous equations with two unknown parameters ($\beta_0 + \beta_1$) and four unknown residuals (r_1, r_2, r_3 and r_4).

Introduction to Matrices

To solve the above equations it is possible to express the values in matrices such that

$$\begin{bmatrix} 10 \\ 15 \\ 20 \\ 23 \end{bmatrix} = \begin{bmatrix} 1 & 1 \\ 1 & 2 \\ 1 & 3 \\ 1 & 4 \end{bmatrix} \begin{bmatrix} \beta_0 \\ \beta_1 \end{bmatrix} + \begin{bmatrix} r_1 \\ r_2 \\ r_3 \\ r_4 \end{bmatrix}$$

$$Y \quad = \quad X \quad \quad \beta \quad + \quad R \qquad (5.3)$$

where Y is a 4 row by 1 column matrix known as the *vector of responses* and contains the responses y_1 to y_4. It is a column vector because it has only one column. If there is only one row and several columns in the matrix it is termed a row vector. R is also a column vector which contains the unknown residuals (r_1–r_4), and is known as the *vector of residuals*. X is the *matrix of parameter coefficients* which contains all the parameter coefficients for the model and has four rows and two columns. It is therefore a 4×2 matrix (rows always given first). It contains a term for an intercept and a straight-line slope. Each element within the matrix of parameter coefficients may be represented by x_{ij} where i is the row and j is the column. The individual elements of X are therefore:

$$x_{11} = 1 \qquad\qquad x_{12} = 1$$
$$x_{21} = 1 \qquad\qquad x_{22} = 2$$
$$x_{31} = 1 \qquad\qquad x_{32} = 3$$
$$x_{41} = 1 \qquad\qquad x_{42} = 4$$

This matrix of parameter coefficients is multiplied by the *matrix of parameters* (β) which contains the parameters the experimenter wishes to estimate. This is a 2×1 column vector containing β_0 and β_1.

There are also some special types of matrices which it might come in handy to know about and I will introduce them as I go along. The first of these is a *scalar*, which put very simply is a single number or a 1×1 matrix.

Π   For the 3×3 matrix A given below, what value is taken by element a_{23}?

$$A = \begin{bmatrix} 4 & 1 & 2 \\ 3 & 5 & 6 \\ 7 & 9 & 8 \end{bmatrix}$$

Taking the second row given by the first subscript (2), only the numbers 3, 5 and 6 are possible. Since the column subscript is always second, it can only be the third column which has the number 6 in it. Therefore $a_{23} = 6$.

5.3. GENERALIZED LEAST-SQUARES MATRIX SOLUTION

Equation 5.3 represents the responses as a set of four simultaneous equations in which there are two unknown parameters, β_0 and β_1, and four unknown residuals, r_1, r_2, r_3 and r_4. Clearly it is impossible to determine these six quantities in four simultaneous equations. An answer is provided by the technique of least squares which minimizes the sum of squares of the residuals and estimates the parameters. I do not intend to prove this here since it would take up quite a bit of space and is not really necessary. However, the generalized matrix solution for a set of b parameters that estimate the β parameters, whilst at the same time minimizing the sum of squares of the residuals, is given by

$$\widehat{B} = (X'X)^{-1}X'Y \qquad (5.4)$$

Here again some of the matrix notation may be unfamiliar. $\widehat{B}$ is a column vector containing the parameters b_0, b_1 etc., which estimate β_0, β_1 etc. These are often called the least squares estimates, denoting that they are derived by this method. X' is called a *transpose* of matrix X. This transpose premultiplies X to give $X'X$. The *inverse* of this matrix is calculated and is represented as $(X'X)^{-1}$. This inverse premultiplies $X'Y$ to give the solution to the estimated parameters.

The next few sections deal with some of the matrix manipulations required to solve this general equation. If you feel comfortable with these manipulations just do the in-text questions to keep up with the example calculations. Alternatively, follow the sub-sections through to refresh your memory.

Transposing a Matrix or Vector

Before premultiplying a matrix by itself or by another matrix it is often necessary to convert it to one of the correct dimensions. It is impossible to multiply a 3×2 matrix by itself because the number of columns in the first matrix is not the same as the number of rows in the second matrix. This will be shown later. For a matrix to be multiplied by itself it is often necessary to find the transpose. The transpose A' of a matrix A is formed by interchanging the rows and columns of A so that element a_{ij} of matrix A becomes element a_{ji} of matrix A'. For example, if

$$A = \begin{bmatrix} 1 & 2 \\ 1 & 4 \\ 1 & 6 \end{bmatrix}$$

then

$$A^t = \begin{bmatrix} 1 & 1 & 1 \\ 2 & 4 & 6 \end{bmatrix}$$

The first column of A becomes the first row of A^t, and the second column of A becomes the second row of A^t. The dimensions of the matrix have been switched around. What was a 3×2 matrix has now been transposed into a 2×3 matrix. If you so wished you could now multiply $A^t{}_{2\times3}$ by $A_{3\times2}$ to form a matrix $A^tA_{2\times2}$.

If the transpose of a square or symmetric matrix is formed then the elements along the principal diagonal running from the top-left corner to the bottom-right corner remain the same. Therefore, given

$$A = \begin{bmatrix} 1 & 4 & 4 \\ 3 & 1 & 7 \\ 3 & 6 & 1 \end{bmatrix}$$

its transpose is

$$A^t = \begin{bmatrix} 1 & 3 & 3 \\ 4 & 1 & 6 \\ 4 & 7 & 1 \end{bmatrix}$$

For a column vector C of n rows by 1 column, the transpose C^t is an n column by 1 row vector. Column vectors are often expressed as their row vectors to save space. Thus if

$$C = \begin{bmatrix} 1 \\ 2 \\ 3 \end{bmatrix} \quad \text{then} \quad C^t = [1 \quad 2 \quad 3]$$

Multiplication of Matrices

In the equation for the generalized matrix solution of simultaneous equations (Eq. 5.4), several of the matrices were multiplied together. The intention here is to show you how to do this. As mentioned earlier, any ordinary number is a scalar or 1×1 matrix and one of the simplest operations in matrix algebra is to multiply a matrix by a scalar. Given a 2×2 matrix and a scalar of 3, the product is found by multiplying each of the elements of the matrix by 3. Thus

$$3 \begin{bmatrix} 5 & 2 \\ 1 & 4 \end{bmatrix} = \begin{bmatrix} 3 \times 5 & 3 \times 2 \\ 3 \times 1 & 3 \times 4 \end{bmatrix} = \begin{bmatrix} 15 & 6 \\ 3 & 12 \end{bmatrix}$$

Multiplying two matrices in which both have dimensions greater than this is rather more tricky. One problem is that two matrices can only be multiplied if the number of columns of the first matrix is equal to the number of rows of the second matrix. If for example, there are two matrices $C_{m \times p}$ and $D_{p \times n}$, the product $E = CD$ is possible because there are p columns in C and p rows in D. The resulting matrix will have m rows and n columns. On the other hand, product DC is only possible if $n = m$.

To calculate the individual elements of the product matrix $E = CD$, you have to multiply each of the p elements in a column of D by the corresponding p elements in a row of C and sum the products. For example, if

$$C = \begin{bmatrix} 1 & 2 \\ 3 & 4 \end{bmatrix} \quad \text{and} \quad D = \begin{bmatrix} 6 & 5 \\ 7 & 8 \end{bmatrix}$$

the first element e_{11} in the product matrix $E = CD$ is calculated thus:

$$\begin{bmatrix} 1 & 2 \\ - & - \end{bmatrix} \begin{bmatrix} 6 & - \\ 7 & - \end{bmatrix} = \begin{bmatrix} (1 \times 6 + 2 \times 7) & - \\ - & - \end{bmatrix} = \begin{bmatrix} 20 & - \\ - & - \end{bmatrix}$$

Similarly to form element e_{21}:

$$\begin{bmatrix} - & - \\ 3 & 4 \end{bmatrix} \begin{bmatrix} 6 & - \\ 7 & - \end{bmatrix} = \begin{bmatrix} - & - \\ 3 \times 6 + 4 \times 7 & - \end{bmatrix} = \begin{bmatrix} - & - \\ 46 & - \end{bmatrix}$$

element e_{12}:

$$\begin{bmatrix} 1 & 2 \\ - & - \end{bmatrix} \begin{bmatrix} - & 5 \\ - & 8 \end{bmatrix} = \begin{bmatrix} - & 1 \times 5 + 2 \times 8 \\ - & - \end{bmatrix} = \begin{bmatrix} - & 21 \\ - & - \end{bmatrix}$$

and finally element e_{22}

$$\begin{bmatrix} - & - \\ 3 & 4 \end{bmatrix} \begin{bmatrix} - & 5 \\ - & 8 \end{bmatrix} = \begin{bmatrix} - & - \\ - & 3 \times 5 + 4 \times 8 \end{bmatrix} = \begin{bmatrix} - & - \\ - & 47 \end{bmatrix}$$

We have therefore calculated all the individual elements in matrix E such that

$$E = \begin{bmatrix} 20 & 21 \\ 46 & 47 \end{bmatrix}$$

∏   Try to find the product $P = DC$ for the above 2×2 matrices D and C. Confirm whether or not P differs from E.

The product $P = DC$ you should have obtained is

$$P = \begin{bmatrix} 21 & 32 \\ 31 & 46 \end{bmatrix}$$

where each element was calculated as follows:

$$P = \begin{bmatrix} (6 \times 1 + 5 \times 3) & (6 \times 2 + 5 \times 4) \\ (7 \times 1 + 8 \times 3) & (7 \times 2 + 8 \times 4) \end{bmatrix}$$

In contrast to ordinary algebra, matrix multiplication is not, in general, commutative. Therefore CD may or may not equal DC. It is, however, distributive, such that $C(D + E) = CD + CE$ and associative, such that $(CD)E = C(DE)$.

∏   Now try a rather easier multiplication involving a 2×1 column vector and a 2×2 matrix. Find the products $P = CD$ and $Q = DC$ given

$$C = \begin{bmatrix} 1 & 6 \\ 1 & 7 \end{bmatrix} \quad \text{and} \quad D = \begin{bmatrix} 3 \\ 2 \end{bmatrix}$$

You should have obtained the product $P = CD$ as follows:

$$P = \begin{bmatrix} (1 \times 3 + 6 \times 2) \\ (1 \times 3 + 7 \times 2) \end{bmatrix} = \begin{bmatrix} 15 \\ 17 \end{bmatrix}$$

It is, however, impossible to obtain a product $Q = DC$ because the first matrix D has only one column which does not equal the number of rows in the second matrix C.

Matrix Inversion

In Eq. (5.4), $\hat{B} = (X'X)^{-1}X'Y$, the superscript -1 is used to indicate that the inverse of matrix $X'X$ has been calculated. Matrix inversion is a fairly complex process which is analogous to the relatively simple process in ordinary algebra or arithmetic of calculating the reciprocal of a number or term. For a scalar this is also true. In a similar way if Z is a square or symmetric matrix (rows = columns), $Z \times Z^{-1} = I$ where I is an *identity matrix* which only has ones as its principal diagonal elements with all other elements being zero. For example, a 2×2 identity matrix I is given below

$$I = \begin{bmatrix} 1 & 0 \\ 0 & 1 \end{bmatrix}$$

An example of an inverse matrix Z^{-1} is

$$Z^{-1} = \begin{bmatrix} 2 & 1 \\ 1 & 2 \end{bmatrix}^{-1} = \begin{bmatrix} 2/3 & -1/3 \\ -1/3 & 2/3 \end{bmatrix}$$

Multiplying the two matrices Z and Z^{-1} together in any order produces an identity matrix:

$$\begin{bmatrix} 2 & 1 \\ 1 & 2 \end{bmatrix} \begin{bmatrix} 2/3 & -1/3 \\ -1/3 & 2/3 \end{bmatrix} = \begin{bmatrix} (4/3 - 1/3) & (-2/3 + 2/3) \\ (2/3 - 2/3) & (-1/3 + 4/3) \end{bmatrix} = \begin{bmatrix} 1 & 0 \\ 0 & 1 \end{bmatrix}$$

For a 2×2 matrix the inverse is easily found (if it exists) by the following method:

Given a matrix Z, its inverse is Z^{-1}

$$\begin{bmatrix} z_{11} & z_{12} \\ z_{21} & z_{22} \end{bmatrix}^{-1} = \begin{bmatrix} z_{22}/D & -z_{12}/D \\ -z_{21}/D & z_{11}/D \end{bmatrix}$$

where D, known as the *determinant*, is found from

$$D = z_{11} \times z_{22} - z_{12} \times z_{21}.$$

It is also usual to express the determinant of a matrix between two vertical lines $|Z|$. Sometimes an inverse cannot be calculated because $D = 0$, with the subsequent division by zero being impossible. A matrix with a determinant of zero is described as being *singular*.

A special case of matrix inversion occurs when the matrix has non-zero elements only on its principal diagonal (running top left to bottom right). Here the inverse is more simply found by taking the reciprocals of each of the diagonal elements. Thus

$$\begin{bmatrix} 4 & 0 & 0 \\ 0 & 3 & 0 \\ 0 & 0 & 2 \end{bmatrix}^{-1} = \begin{bmatrix} 1/4 & 0 & 0 \\ 0 & 1/3 & 0 \\ 0 & 0 & 1/2 \end{bmatrix}$$

These types of inverse matrix are often found in experimental designs.

Π   Suppose that a 2 × 2 matrix A was $\begin{bmatrix} 2 & 2 \\ 1 & 3 \end{bmatrix}$

What is its inverse A^{-1}?

To obtain the inverse of A you should first have calculated the determinant D by subtracting the products of the diagonally opposite elements such that $D = (a_{11} \times a_{22}) - (a_{12} \times a_{21})$. The determinant of A is therefore $(2 \times 3 - 1 \times 2) = 4$. The inverse of A is then given by:

$$A^{-1} = \begin{bmatrix} a_{22}/D & -a_{12}/D \\ -a_{21}/D & -a_{11}/D \end{bmatrix} = \begin{bmatrix} 3/4 & -2/4 \\ -1/4 & 2/4 \end{bmatrix}$$

5.3.1. Calculation of the Estimated Parameters

Applying the matrix manipulations you have just learned to solve the generalized least squares equation (5.4) should enable you to obtain some estimates of β_0 and β_1 for the single-factor first-order example. To recap then, we have a matrix of parameter coefficients X, and a vector of responses Y.

$$\begin{bmatrix} 10 \\ 15 \\ 20 \\ 23 \end{bmatrix} = \begin{bmatrix} 1 & 1 \\ 1 & 2 \\ 1 & 3 \\ 1 & 4 \end{bmatrix} \begin{bmatrix} \beta_0 \\ \beta_1 \end{bmatrix} + \begin{bmatrix} r_1 \\ r_2 \\ r_3 \\ r_4 \end{bmatrix}$$

$$Y \quad = \quad X \quad \beta \quad + \quad R$$

The transpose of X is $X^t = \begin{bmatrix} 1 & 1 & 1 & 1 \\ 1 & 2 & 3 & 4 \end{bmatrix}$

which premultiplies X to form the matrix X^tX, and Y to form X^tY:

$$X^tX = \begin{bmatrix} 1 & 1 & 1 & 1 \\ 1 & 2 & 3 & 4 \end{bmatrix} \begin{bmatrix} 1 & 1 \\ 1 & 2 \\ 1 & 3 \\ 1 & 4 \end{bmatrix} = \begin{bmatrix} 4 & 10 \\ 10 & 30 \end{bmatrix}$$

$$X^tY = \begin{bmatrix} 1 & 1 & 1 & 1 \\ 1 & 2 & 3 & 4 \end{bmatrix} \begin{bmatrix} 10 \\ 15 \\ 20 \\ 23 \end{bmatrix} = \begin{bmatrix} 68 \\ 192 \end{bmatrix}$$

Applying the technique given above in Section 5.3 for finding the $(X^tX)^{-1}$ inverse, the determinant D of X^tX is found initially, $D = (4 \times 30 - 10 \times 10) = 20$. The inverse of X^tX is therefore

$$(X^tX)^{-1} = \begin{bmatrix} 30/20 & -10/20 \\ -10/20 & 4/20 \end{bmatrix} = \begin{bmatrix} 1.5 & -0.5 \\ -0.5 & 0.2 \end{bmatrix}$$

The estimated b parameters can now be found in one final multiplication step:

$$\hat{B} = (X^tX)^{-1} \qquad X^tY$$

$$\hat{B} = \begin{bmatrix} 1.5 & -0.5 \\ -0.5 & 0.2 \end{bmatrix} \begin{bmatrix} 68 \\ 192 \end{bmatrix} = \begin{bmatrix} 6.0 \\ 4.4 \end{bmatrix} \begin{matrix} = & b_0 \\ = & b_1 \end{matrix}$$

The estimated intercept b_0 is 6.0 and the estimated slope b_1 is 4.4. This has been plotted in Fig. 5.3 to show how well the model fits the data.

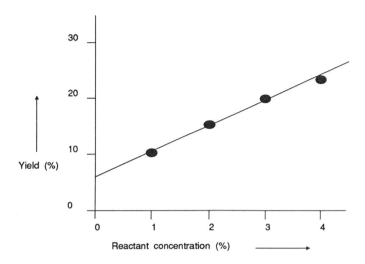

Fig. 5.3. *Plot of yield against reactant concentration fitted to first-order model*

Clearly this is not a perfect fit because some of the points do not lie on the line.

5.3.2. Calculation of Residual Errors

It is possible to calculate the values of the residuals by substituting the estimated parameter values into Eq. (5.3) so that

$$R = Y - X\hat{B}$$

Multiplying the matrix of parameter coefficients by the vector of estimated parameters gives the predicted responses, which can then be subtracted from the observed responses to give a vector of residuals. So far I have not dealt with the subtraction and addition of matrices. Therefore, you may not be able to calculate the residuals. If that is the case, please go through the next section. If, on the other hand, you are reasonably happy with these manipulations just try the in-text questions to keep up with the example calculations.

Addition and Subtraction of Matrices

The sum of two matrices is simply obtained by adding the elements of one matrix to the corresponding elements in another matrix. An important point here is that the matrices being summed should be of the same dimensions. The new matrix will then have the same dimensions as the original matrices. For example, given that

$$E = \begin{bmatrix} 1 & 2 \\ 3 & 4 \end{bmatrix} \quad \text{and} \quad F = \begin{bmatrix} 5 & 6 \\ 7 & 8 \end{bmatrix}$$

and you wish to calculate $A = E + F$.

$$A = \begin{bmatrix} 1 & 2 \\ 3 & 4 \end{bmatrix} + \begin{bmatrix} 5 & 6 \\ 7 & 8 \end{bmatrix} = \begin{bmatrix} 1+5 & 2+6 \\ 3+7 & 4+8 \end{bmatrix} = \begin{bmatrix} 6 & 8 \\ 10 & 12 \end{bmatrix}$$

If, on the other hand, you had a matrix Z, where

$$Z = \begin{bmatrix} 1 & 2 \\ 3 & 4 \\ 5 & 6 \end{bmatrix}$$

you could not add this matrix to either E or F because the dimensions are not the same. With matrices the order of addition does not matter and is said to be commutative.

$$E + F = F + E$$

Adding F and E would give the same result as the A matrix calculated above.

To obtain the difference between two matrices simply subtract the element in the second matrix from its corresponding elements in the first matrix. For matrices E and F given above, $S = F - E$. Therefore

$$S = \begin{bmatrix} 5 & 6 \\ 7 & 8 \end{bmatrix} - \begin{bmatrix} 1 & 2 \\ 3 & 4 \end{bmatrix} = \begin{bmatrix} 5-1 & 6-2 \\ 7-3 & 8-4 \end{bmatrix} = \begin{bmatrix} 4 & 4 \\ 4 & 4 \end{bmatrix}$$

∏   Calculate the residuals for the example given earlier using

$$R = Y - X\widehat{B}$$

The vector of residuals (R) you should have obtained is $R^t =$ $[-0.4, 0.20, 0.80, -0.60]$ in its transposed form. The way I obtained this vector was first to calculate the estimated responses at the different factor levels by multiplying the X matrix by the estimated parameter matrix. Thus

$$\hat{Y} = \begin{bmatrix} 1 & 1 \\ 1 & 2 \\ 1 & 3 \\ 1 & 4 \end{bmatrix} \begin{bmatrix} 6.0 \\ 4.4 \end{bmatrix} = \begin{bmatrix} (1 \times 6.0) + (1 \times 4.4) \\ (1 \times 6.0) + (2 \times 4.4) \\ (1 \times 6.0) + (3 \times 4.4) \\ (1 \times 6.0) + (4 \times 4.4) \end{bmatrix} = \begin{bmatrix} 10.4 \\ 14.8 \\ 19.2 \\ 23.6 \end{bmatrix}$$

$$\hat{Y} = \quad X \quad \hat{B}$$

Subtracting $\hat{Y}$ from Y gives the residuals R:

$$\begin{bmatrix} 10 \\ 15 \\ 20 \\ 23 \end{bmatrix} - \begin{bmatrix} 10.4 \\ 14.8 \\ 19.2 \\ 23.6 \end{bmatrix} = \begin{bmatrix} -0.4 \\ 0.2 \\ 0.8 \\ -0.6 \end{bmatrix} \begin{matrix} = r_1 \\ = r_2 \\ = r_3 \\ = r_4 \end{matrix}$$

$$Y \quad - \quad \hat{Y} \quad = \quad R$$

In terms of the responses and the fitted straight line these residuals are the actual vertical deviations of the observed responses from the calculated straight line graph. Note that the sum of the residuals in this case is zero. This is generally true for models containing a β_0 term, whereas in models without β_0 terms, the residuals will not necessarily sum to zero.

Earlier I stated that the generalized least-squares matrix equation could be applied to estimate β_0 and β_1 whilst minimizing the sum of squares of the residuals. Having obtained the residuals for this example, we can now calculate the sum of squares of the residuals *(ResidSS)* as $R^t R$.

$$R^t R = [-0.4 \quad 0.2 \quad 0.8 \quad -0.6] \begin{bmatrix} -0.4 \\ 0.2 \\ 0.8 \\ -0.6 \end{bmatrix} = [1.2]$$

We will use this value later to test the adequacy of the model.

5.3.3. Testing the Adequacy of Response-surface Models

In the earlier parts of this book, I introduced analysis of variance (*ANOVA*) to test for the significance of factor effects. It is equally applicable to response surface models because, in all *ANOVA* calculations, the total sum of squares may be partitioned into sums of squares terms associated with factor effects, residual error, correction for the mean and so on.

We can partition the crude total sum of squares for the first-order single-factor example given above into the total sum of squares of the responses corrected for the mean (*TotalSS*) and the sum of squares due to the mean. We can further divide the *TotalSS* into the sum of squares due to the treatment or factor (*TreatmentSS*) and the residual error sum of squares (*ResidSS*). These sums of squares terms are then divided by their degrees of freedom to give the mean squares.

If we continue to use matrix algebra for this, the *TotalSS* is perhaps most easily calculated by subtracting the average response from each of the observed responses, so forming a matrix T where

$$T = \begin{bmatrix} y_1 \\ y_2 \\ y_3 \\ y_4 \end{bmatrix} - \begin{bmatrix} \bar{y} \\ \bar{y} \\ \bar{y} \\ \bar{y} \end{bmatrix} = \begin{bmatrix} 10 \\ 15 \\ 20 \\ 23 \end{bmatrix} - \begin{bmatrix} 17 \\ 17 \\ 17 \\ 17 \end{bmatrix} = \begin{bmatrix} -7 \\ -2 \\ 3 \\ 6 \end{bmatrix}$$

$$T = Y - \bar{Y}$$

We then calculate the total sum of squares corrected for the mean by pre-multiplying T by its transpose T'.

$$T'T = \begin{bmatrix} -7 & -2 & 3 & 6 \end{bmatrix} \begin{bmatrix} -7 \\ -2 \\ 3 \\ 6 \end{bmatrix} = [98]$$

We can now calculate the sum of squares due to the factor(s), which we have also called the sum of squares due to the treatments, by subtracting the average response from the estimated responses, so obtaining a matrix due to factor effects, F:

$$F = \hat{Y} - \bar{Y} \quad \text{where} \quad \hat{Y} = X\hat{B}$$

$$F = \begin{bmatrix} 10.4 \\ 14.8 \\ 19.2 \\ 23.6 \end{bmatrix} - \begin{bmatrix} 17 \\ 17 \\ 17 \\ 17 \end{bmatrix} = \begin{bmatrix} -6.6 \\ -2.2 \\ 2.2 \\ 6.6 \end{bmatrix}$$

which can then supply the sum of squares due to the factor (*TreatmentSS*) by pre-multiplying F by its transpose F^t.

$$F^t F = \begin{bmatrix} -6.6 & -2.2 & 2.2 & 6.6 \end{bmatrix} \begin{bmatrix} -6.6 \\ -2.2 \\ 2.2 \\ 6.6 \end{bmatrix} = \begin{bmatrix} 96.8 \end{bmatrix}$$

We could just have easily calculated this by subtracting the residual sum of squares from the total sum of squares corrected for the mean.

Π Complete the analysis of variance table given below for this example and decide whether the model is significant at the $P = 0.05$ probability level.

Source of variation	d.f.	SS	MS	Variance ratio
Total (corrected)		98		
Treatment		96.8		
Residual error		1.2		

Since the sums of squares are provided, all you had to decide was the number of degrees of freedom associated with each source of variance. Since there are four experimental runs, only three *d.f.* are now available because the total sum of squares has been corrected for the mean. I hope you realised that it contains only one effect, that due to the regression line or slope (b_1). Hence *TreatmentSS* has only one *d.f.*, leaving two *d.f.* for the residual sum of squares, *ResidSS*. You should therefore have divided the *ResidSS* by 2 to obtain a residual error mean square of 0.6. You should then have had two mean square values, which are by definition estimates of the variance of the results. These should be similar if the regression does not account for a significant portion of the total variance. Working on the null hypothesis that they are of similar magnitude, you should have calculated a variance ratio of *TreatmentMS/ResidMS* to complete the *ANOVA* table given below.

Source of variation	*d.f.*	*SS*	*MS*	Variance ratio (F)
Total (corrected)	3	98		
Treatment	1	96.8	96.8	161.33
Residual	2	1.2	0.6	

Having done this you should have looked up an F value for one and two *d.f.* at a probability level of 0.05 (18.51) which has to be exceeded to reject the null hypothesis. This is clearly the case. You should then have accepted an alternative hypothesis that the regression accounts for a significant portion of the variance.

Here are two SAQs to enable you to test your understanding. The responses to them begin on page 260.

SAQ 5.3a

The following statements refer to different models and hypotheses. Indicate whether the statements are true (T) or false (F).

(i) In the model $y_i = \beta_1 x_{1i} + r_i$ there will be no intercept and the data will be forced through zero. [True/False]

(ii) When the model fits the data perfectly, estimates of the predicted responses given by the matrix equation

$$\hat{Y} = X\hat{B}$$

will reproduce the original responses. [True/False]

(iii) In the model $y_i = 0 + r_i$, the experimenter wishes to test the hypothesis that the intercept is significantly different from zero.[True/False]

(iv) When the model $y_i = \beta_0 + \beta_1 x_{1i}$ is used for a number of experimental runs all at the same level of x_1, the intercept can be zero. [True/False]

(v) The experimenter can be at least 95% confident in rejecting the null hypothesis if the calculated value of F (variance ratio) exceeds the tabulated value at the $P = 0.05$ probability level. [True/False]

SAQ 5.3b

In an experiment to investigate the effect of pH upon the concentration of a complex species, the pH levels used were 2.0, 3.0 and 4.0, and the observed concentrations of a species Y were 0.30, 0.46 and 0.59 (mol dm^{-3}). Determine the parameters for a first-order model incorporating the slope and intercept, and decide by means of an analysis of variance whether the model can be accepted as a reasonable fit of the data at a probability level of $P = 0.05$.

5.4. SINGLE-FACTOR SECOND-ORDER MODEL

In the previous section, the model used was first-order, yielding estimates of the intercept b_0, and the straight-line slope b_1. It is, however, possible to change the model and subsequent analysis of the results to include a second-order parameter for curvature, given by β_{11}, and estimated as b_{11}. The model is now

$$y_i = \beta_0 + \beta_1 x_{1i} + \beta_{11} x_{1i}^2 + r_i \tag{5.5}$$

The matrix of parameter coefficients has an additional third column (x_1^2) composed of the squares of the values in the second column;

$$X = \begin{bmatrix} x_0 & x_1 & x_1^2 \\ 1 & 1 & 1 \\ 1 & 2 & 4 \\ 1 & 3 & 9 \\ 1 & 4 & 16 \end{bmatrix}$$

The matrix model is the same as used previously, $Y = X\hat{B} + R$, and can be solved using Eq. (5.4). If, however, you try to calculate $X'X$ and find its inverse you might encounter some difficulties because $X'X$ is a 3×3 matrix instead of a 2×2 matrix. If you can calculate these inverses move on to try the in-text question. Otherwise follow the next sub-section through.

Inversion of 3×3 Matrices

Suppose that we require the inverse A^{-1} of matrix A given below, in order to estimate some parameters. As with 2×2 matrices, you have to calculate the determinants, and a procedure is required to do this.

$$A = \begin{bmatrix} a_{11} & a_{12} & a_{13} \\ a_{21} & a_{22} & a_{23} \\ a_{31} & a_{32} & a_{33} \end{bmatrix} = \begin{bmatrix} 1 & 1 & 1 \\ 1 & 2 & 3 \\ 1 & 3 & 6 \end{bmatrix}$$

If we select element a_{11} which is in row 1 and column 1, then the elements not in its row and column are

$$\begin{bmatrix} a_{22} & a_{23} \\ a_{32} & a_{33} \end{bmatrix}$$

This 2×2 structure is called a minor of element a_{11}. We can calculate the determinant of this 2×2 minor in the same way as for other 2×2 matrices and it is given by $(a_{22}a_{33}) - (a_{32}a_{23})$. Similarly the minor of a_{12} is

$$\begin{bmatrix} a_{21} & a_{23} \\ a_{31} & a_{33} \end{bmatrix}$$

and its determinant is given by $(a_{21}a_{33}) - (a_{31}a_{23})$.

The minor of element a_{13} is

$$\begin{bmatrix} a_{21} & a_{22} \\ a_{31} & a_{32} \end{bmatrix}$$

and its determinant is given by $(a_{21}a_{32}) - (a_{31}a_{22})$.

However, there is a notional sign of either $+$ or $-$ associated with each element in the 3×3 matrix. These are shown below:

$$\begin{bmatrix} + & - & + \\ - & + & - \\ + & - & + \end{bmatrix}$$

When we calculate the determinant of the minor of a given element, we have to associate with the determinant the sign associated with the appropriate element. These signed determinants are then called cofactors.

The cofactor of element a_{11} is $+(a_{22}a_{33} - a_{32}a_{23}) = A_{11}$, and the cofactor of element a_{12} is $-(a_{21}a_{33} - a_{23}a_{31}) = A_{12}$. The determinant of the 3×3 matrix is given by the sum of the products of any row or column elements and their appropriate cofactors. Since the cofactors for the first row of the 3×3 matrix have already been identified, the determinant of the 3×3 matrix is given by

$$|A| = a_{11}A_{11} + a_{12}A_{12} + a_{13}A_{13}$$

or expressed in long-hand

$$|A| = a_{11}(a_{22}a_{33} - a_{32}a_{23}) - a_{12}(a_{21}a_{33} - a_{23}a_{31}) + a_{13}(a_{21}a_{32} - a_{22}a_{31})$$

$$= 1 \times (2 \times 6 - 3 \times 3) - 1 \times (1 \times 6 - 1 \times 3) + 1 \times (1 \times 3 - 2 \times 1) = 1$$

Once we have the determinant, we find the inverse by dividing the cofactors of each element by the determinant. However, we first of all transpose the matrix of cofactors as shown in the following matrix:

$$\begin{bmatrix} A_{11}/D & A_{21}/D & A_{31}/D \\ A_{12}/D & A_{22}/D & A_{32}/D \\ A_{13}/D & A_{23}/D & A_{33}/D \end{bmatrix}$$

Therefore A^{-1} is

$$A^{-1} = \begin{bmatrix} (2 \times 6 - 3 \times 3)/1 & -(1 \times 6 - 1 \times 3)/1 & (1 \times 3 - 2 \times 1)/1 \\ -(1 \times 6 - 1 \times 3)/1 & (1 \times 6 - 1 \times 1)/1 & -(1 \times 3 - 1 \times 1)/1 \\ (1 \times 3 - 2 \times 1)/1 & -(1 \times 3 - 1 \times 1)/1 & (1 \times 2 - 1 \times 1)/1 \end{bmatrix}$$

$$A^{-1} = \begin{bmatrix} 3 & -3 & 1 \\ -3 & 5 & -2 \\ 1 & -2 & 1 \end{bmatrix}$$

The inversion of most square matrices larger than 3×3 is possible by hand, but is usually best left to computers because of the number of calculations involved. Only if they fall into the special case of having non-zero only elements on the principal diagonal can they easily be inverted by finding the reciprocals of the individual elements.

If you found you could do the above matrix inversion without too much difficulty, move on to the SAQ given below. Otherwise go through the example again and then try the in-text question.

∏   Find the inverse Z^{-1} of the following 3×3 matrix Z.

$$Z = \begin{bmatrix} 3 & 2 & 1 \\ 2 & 4 & 2 \\ 1 & 8 & 3 \end{bmatrix}$$

The determinant you should have obtained is 4. Taking any row or column it is possible to work out the cofactors of the elements contained therein. For the top row, the determinant of minor z_{11} is $(4 \times 3 - 2 \times 8) = -4$, for the minor of element z_{12}, the determinant is $(2 \times 3) - (2 \times 1) = 4$ and for the minor of element z_{13}, the determinant is $(2 \times 8) - (4 \times 1) = 12$. Applying the signs +, − and + respectively to obtain the cofactors and summing, you should have obtained the answer −8.

With this you could have calculated the inverse Z^{-1} by dividing a matrix of the transposed cofactors by the determinant. Thus, Z^{-1} is

$$Z^{-1} = \begin{bmatrix} (4 \times 3 - 2 \times 8)/-8 & -(2 \times 3 - 8 \times 1)/-8 & (2 \times 2 - 1 \times 4)/-8 \\ -(2 \times 3 - 2 \times 1)/-8 & (3 \times 3 - 1 \times 1)/-8 & -(3 \times 2 - 2 \times 1)/-8 \\ (2 \times 8 - 4 \times 1)/-8 & -(3 \times 8 - 2 \times 1)/-8 & (3 \times 4 - 2 \times 2)/-8 \end{bmatrix}$$

$$Z^{-1} = \begin{bmatrix} 0.50 & -0.25 & 0.00 \\ 0.50 & -1.00 & 0.50 \\ -1.50 & +2.75 & -1.00 \end{bmatrix}$$

SAQ 5.4a

Determine the least-squares estimates for the intercept (b_0), slope (b_1) and curvature (b_{11}) in the single-factor example given its $X'X$ matrix made up from the matrix of parameter coefficients X, containing a column of 1's for the intercept, a column containing the values 1, 2, 3 and 4 for the levels used, and a column containing the values 1, 4, 9 and 16 for the squares of the coefficients.

$$X'X = \begin{bmatrix} 1 & 1 & 1 & 1 \\ 1 & 2 & 3 & 4 \\ 1 & 4 & 9 & 16 \end{bmatrix} \begin{bmatrix} 1 & 1 & 1 \\ 1 & 2 & 4 \\ 1 & 3 & 9 \\ 1 & 4 & 16 \end{bmatrix}$$

$$= \begin{bmatrix} 4 & 10 & 30 \\ 10 & 31 & 100 \\ 30 & 100 & 354 \end{bmatrix}$$

SAQ 5.4a

5.4.1. Coefficients of Determination and Correlation

Analysis of variance is one way of deciding whether the model used is an adequate fit to the data for the response surface. When the model has a β_0 term, the *ResidSS* and *TreatmentSS* together add up to the total (corrected) sum of squares, *TotalSS*. When *TreatmentSS* approaches *TotalSS*, most of the variance in the responses is caused by the factors, whereas when *ResidSS* approaches *TotalSS*, little of the variance is the result of the effects of the factors.

A ratio of *TreatmentSS* to *TotalSS* is called the *coefficient of multiple determination* (R^2), and varies from 0, where the factors do not explain the variance in the data, up to a value of 1, the point at which the factors perfectly explain the data. Its square root (R) is the *coefficient of multiple correlation*. For the case of a single-factor first-order model (straight line) $y_i = \beta_0 + \beta_1 x_{1i} + r_i$, R^2 can be used to supply the *correlation coefficient* (r), where $r = \text{sign}\ (b_1)\sqrt{(R^2)}$. The sign is used to describe the form of relationship between y and x_1, with a negative slope indicating that the response y decreases as x_1 increases, and a positive sign indicating that as x_1 increases so does y. A correlation coefficient varying between -1 and $+1$ indicates therefore how closely the factor explains the data.

$\prod$ What percentage of the total (corrected) sum of squares is explained by the factor when an R value of 0.90 is reported?

Only 81% of the total (corrected) sum of squares is accounted for since R^2 is the ratio of *TreatmentSS* to *TotalSS*. Thus $0.9^2 = 0.81$. This illustrates a common misconception when quoting the coefficient of multiple correlation, R, or correlation coefficient, r. An apparently high R or r value can give an impression of a good fit of the model which is not necessarily true. Also, by adding more terms to the model it might be possible to increase R and R^2 without necessarily having a significantly better fit. These coefficients do not take account of the shift of the degrees of freedom from the residual sum of squares to the factor or regression sum of squares, which can indicate whether a better fit has been obtained. Therefore R and R^2 should not be used by themselves to indicate the effectiveness of factors.

SAQ 5.4b	For the single-factor example, calculate the coefficient of multiple correlation (R^2) and correlation coefficient (r), and compare these with the F ratios calculated previously. Does the second-order model provide a better fit than the first-order model?

5.4.2. Replication

If some of the experimental runs in a design are replicated, that is, they have the same set of factor levels, the responses from these replicates can be used to provide a mean response and an estimate of pure experimental uncertainty, σ^2_{pe}, at that set of factor levels, as given by

$$s^2_{pe} = \frac{\Sigma(\bar{y} - y_i)^2}{n - 1} \tag{5.6}$$

where $\bar{y}$ and y_i refer to the n replicates at the same factor levels i.e., have the same treatments. For example, given the following five experimental runs, three of which are replicates,

Factor level	Response
2	3
4	6
3	4.6
3	4.5
3	4.4

the estimate s^2_{pe} is given by:

$$s^2_{pe} = \frac{[(4.5 - 4.6)^2 + (4.5 - 4.5)^2 + (4.5 - 4.4)^2]}{3 - 1}$$

$$= 0.02/2 = 0.01$$

If the factor has no effect upon the response then a model

$$y_i = \beta_0 + r_i$$

may be applied, and the residual variance s^2_r, calculated from the variation of all the responses about the average response, should provide an unbiased estimate of pure experimental uncertainty, σ^2_{pe}.

$$s^2_r = \frac{[(4.5 - 3)^2 + (4.5 - 6)^2 + (4.5 - 4.6)^2 + (4.5 - 4.5)^2 + (4.5 - 4.4)^2]}{5 - 1}$$

$$= 4.52/4 = 1.13$$

However, it is obvious that the estimates are not the same and that this model is inadequate. It is possible to test this statistically with an F test since the residual sum of squares can be shown to be composed of the *Sum of Squares due to Purely Experimental Uncertainty* (*PureSS*) and the *Sum of Squares due to Lack of Fit* (*LofSS*) such that

$$ResidSS = LofSS + PureSS \qquad (5.7)$$

This *LofSS* divided by the appropriate degrees of freedom also provides an estimate of the variance which, if the model is correct, provides an unbiased estimate of σ^2_{pe}. If, on the other hand, the model is inappropriate, *LofSS* will be inflated and the variance estimate, s^2_{lof}, will be biased. As well as an additivity of sums of squares, the degrees of freedom are also additive such that

$$Resid\ d.f. = Lof\ d.f. + Pure\ d.f. \qquad (5.8)$$

In the above example therefore,

$$LofSS = ResidSS - PureSS = 4.52 - 0.02 = 4.50.$$

$$s^2_{lof} = 4.50/2 = 2.25$$

The estimated variance due to purely experimental uncertainty was calculated earlier:

$$s^2_{pe} = 0.01$$

The variance ratio between these two quantities may now be calculated:

$$F_{(lof/pe)} = 2.25/0.01 = 225$$

From the table of F values given in Fig. 1.3b, it can be seen that this calculated variance ratio, $F_{(lof/pe)}$ of 225 with 2, 2 $d.f.$ is significant at the

$P = 0.05$ probability level, thus confirming our suspicion that the model is inadequate. Applying a different model $y_i = \beta_0 + \beta_1 x_{1i} + r_i$ in which the factor plays some part, the parameter estimates work out to be $b_0 = 0$ and $b_1 = 1.5$ with a residual vector:

$$R^t = [0 \quad 0 \quad 0.1 \quad 0 \quad -0.1]$$

The last three of these residuals give the same estimate of variance due to pure experimental uncertainty as before

$$s^2_{pe} = \left[0.1^2 + 0^2 + -0.1^2\right] / 2 = 0.02/2 = 0.01$$

However, since the first two values are zero, the *ResidSS* also takes the same value, that is to say there is no lack of fit in the model, rendering an *F*-test unnecessary.

5.4.3. Variances of Estimated Parameters

If a response surface experiment is repeated under what should be exactly the same conditions, it is unlikely that the same responses would be obtained on both occasions. As a consequence the estimates for the parameters β_0, β_1 etc. will be different. They are, however, likely to vary within certain limits which we can estimate. This may be of interest when we are trying to decide whether a parameter is significantly different from zero at a certain probability level, for example, the β_0 parameter in a model $y_i = \beta_0 + r_i$. One way we can do this is to calculate a confidence interval for the parameter to determine whether it includes zero, as in Fig. 5.4b. Alternatively, we could work out the probability level required for the confidence interval of b_0 to include zero.

The usual way of calculating these confidence intervals is to check first of all that the residuals are distributed normally and that no trends are apparent. The parameter estimates obtained will usually then vary in normal manner. We can calculate a variance for each parameter, e.g. s^2_{b0}, by multiplying each element of the $(X'X)^{-1}$ matrix by the estimate of pure experimental uncertainty to form a matrix V, known as the variance–covariance matrix. Along the principal diagonal of V from the upper left corner to the lower right corner are the variances of the parameter estimates. Each off-diagonal value is a covariance between two parameter estimates. If $X'X$ is a 2×2

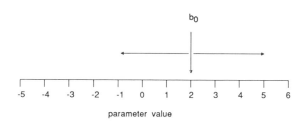

Fig. 5.4b.   *Confidence interval for estimated parameter b_0 which includes
zero*

matrix calculated on the basis of a model $y_i = \beta_0 + \beta_1 x_{1i} + r_i$, the top left-hand
element of V is $s_{b_0}^2$ and the bottom right-hand element is $s_{b_1}^2$.

For the example given above in Section 5.4.2. the X matrix is

$$X = \begin{bmatrix} x_0 & x_1 \\ 1 & 2 \\ 1 & 4 \\ 1 & 3 \\ 1 & 3 \\ 1 & 3 \end{bmatrix} \quad \text{and} \quad X^tX = \begin{bmatrix} 5 & 15 \\ 15 & 47 \end{bmatrix}$$

The determinant of X^tX is $(5 \times 47 - 15 \times 15) = 10$ from which $(X^tX)^{-1}$ can
be calculated as before:

$$(X^tX)^{-1} = \begin{bmatrix} 4.70 & -1.50 \\ -1.50 & 0.50 \end{bmatrix}$$

The estimate of the variance of pure experimental uncertainty, $s^2{}_{pe}$, was
calculated to be 0.01. Therefore,

$$V = \begin{bmatrix} 0.047 & -0.015 \\ -0.015 & 0.005 \end{bmatrix}$$

The variance of the estimate b_0 is $s_{b_0}^2 = 0.047$. The confidence interval (*CI*
for b_0 is calculated by multiplying a t value with the appropriate degrees of
freedom (at a given probability level) by this value of s_{b0}. The t value for
two *d.f.* at a probability level of $P = 0.05$ is 4.303 and $s_{b0} = \sqrt{0.047} = 0.217$.
Therefore, $CI_{b0} = b_0 \pm 0.217 \times 4.303 = b_0 \pm 0.933$.

Π   What is the confidence interval for the estimate b_1?

From the variance–covariance matrix, V, given above, $s_{b1} = \sqrt{0.005} = 0.0707$. Using the same t value, I calculated the confidence interval of b_1 as $CI_{b1} = b_1 \pm 0.0707 \times 4.303 = b_1 \pm 0.304$. I calculated the value of b_1 to be 1.50 which could therefore be expected to vary between 1.2 and 1.8.

Obviously we wish to have small confidence intervals about the estimates of the parameters and, since the $(X'X)^{-1}$ matrix has an influence on these confidence intervals, minimizing the elements contained within this matrix might improve the design.

SAQ 5.4c

Consider two designs based on the model $y_i = \beta_0 + \beta_1 x_{1i} + r_i$ in which only two levels are used, the first level being the same for both designs and the second level being different, such that the X matrices are as given below.

Design 1

$$X_1 = \begin{bmatrix} 1 & 1 \\ 1 & 3 \end{bmatrix}$$

Design 2

$$X_2 = \begin{bmatrix} 1 & 1 \\ 1 & 10 \end{bmatrix}$$

What effect does changing the level of the second run have upon the estimates of uncertainty in the parameters? How does this influence the choice of experimental design if the experimenter wants as small as possible an estimate of uncertainty for b_1?

SAQ 5.4c

5.4.4. Coding of Factor Levels

It is not only the variances of the parameters which can be affected by the experimental design, but also the covariances (off-diagonal elements of V), as shown above. For the example used to examine the importance of replication (Sections 5.4.2 and 5.4.3), the estimate of covariance between b_0 and b_1 [$s^2_{(b_0 b_1)}$] took a value of -0.015. This covariance is an indication of the extent to which the estimates of the parameters b_0 and b_1 depend on one another, a positive sign indicating that as b_0 increases then so does b_1, and a negative sign indicating that as b_0 increases, so b_1 decreases. We want covariances between parameter estimates to be as small as possible, and one way of accomplishing this is to code the factor levels about the central point which is the level taken by runs 3, 4 and 5. This is converted to a coded value of zero. We then normalize the interval between the levels to take values of unity. Any design may be coded in this way using

$$x_{1i}* \;=\; (x_{1i} \;-\; c_{x1})/d_{x1} \tag{5.9}$$

in which $x_{1i}*$ is the coded value of factor x_1 in the ith run, c_{x1} is the location of the centre point for factor x_1, and d_{x1} is the interval between the experimental units along factor x_1. The matrix of parameter coefficients for the example is now given by X_*, where

$$X* = \begin{bmatrix} x_0 & x_1 \\ 1 & -1 \\ 1 & 1 \\ 1 & 0 \\ 1 & 0 \\ 1 & 0 \end{bmatrix} \quad \text{and} \quad X*'X* = \begin{bmatrix} 5 & 0 \\ 0 & 2 \end{bmatrix}$$

The determinant of $X*'X*$ is unaffected by the process of coding and remains at $(5 \times 2 - 0 \times 0) = 10$. Hence we can calculate $(X*'X*)^{-1}$:

$$(X*'X*)^{-1} = \begin{bmatrix} 0.20 & 0 \\ 0 & 0.50 \end{bmatrix}$$

$$X*'Y = \begin{bmatrix} 1 & 1 & 1 & 1 & 1 \\ -1 & 1 & 0 & 0 & 0 \end{bmatrix} \begin{bmatrix} 3.0 \\ 6.0 \\ 4.6 \\ 4.5 \\ 4.4 \end{bmatrix} = \begin{bmatrix} 22.5 \\ 3.0 \end{bmatrix}$$

We can now estimate the parameter vector $\widehat{B}*$ in the usual way using the general matrix solution (Eq. 5.4):

$$\widehat{B}* = (X*'X*)^{-1} \quad X*'Y$$

$$\widehat{B}* = \begin{bmatrix} 0.20 & 0 \\ 0 & 0.50 \end{bmatrix} \begin{bmatrix} 22.5 \\ 3.0 \end{bmatrix} = \begin{bmatrix} 4.5 \\ 1.5 \end{bmatrix}$$

The b_1* parameter estimated using the coded coefficients turns out to be the same as when we used the uncoded coefficients (Section 5.4.2), but b_0 and b_0* are substantially different. The covariances between the parameter estimates are now zero. Note also that the variance of the b_1 estimate is unaffected by the process of coding since the bottom right-hand element takes the same value in both the uncoded and coded $(X'X)^{-1}$ matrices. There is an added advantage that the matrix manipulations are also much easier to handle for coded designs as the multiplication products are much smaller. However, the parameter estimates are now different to what they would be in the uncoded designs. For example, b_0* is the estimated response at the centre of the design and b_0 is the estimated response at the origin of the uncentred design. As in the above example these two values are usually quite different, but it is possible to transform coded estimated parameters to what they would be in the original factor space. Since the b_1* and b_1

estimates are the same in this case there is no need to transform the values. However, as we shall see later, you might sometimes wish to know the b_0 value.

For any single-factor model it is easy to transform coded parameter estimates back to their uncoded values by arithmetical means. However, when the models fitted to the data are more complex, e.g. they include quadratic and interaction terms, the estimated parameters often turn out to be substantially different and a method to convert coded estimates to their uncoded estimates is required. A general method of producing uncoded parameter estimates is given by a matrix equation where

$$\hat{B} = T^{-1}\hat{B}_*$$
(5.10)

in which T is a coding matrix given by

$$T = (X_*^t X_*)^{-1} (X_*^t X)$$
(5.11)

and X and X_* represent the uncoded and coded matrix of parameter coefficients respectively. The conversion of coded estimates to their uncoded counterparts is then a matter of manipulating the matrices. We will apply these matrix equations to greater advantage in later sections.

Learning Objectives

After reading the material in Sections 5.1 to 5.4 you should now be able to:

- recognize the applicability of different models to different experimental situations;

- calculate estimates of parameters for single-factor models using the generalized least-squares matrix solution;

- test the adequacy of single-factor models by *ANOVA* techniques;

- understand the effects of the chosen design upon the variances and covariances of the estimated parameters.

5.5. TWO-FACTOR RESPONSE SURFACE DESIGNS

In previous sections of Part 5 the effect of one factor upon a response has been examined and single-factor response surfaces developed. However, as shown in Parts 3 and 4, there is often a joint dependence of the response upon the levels of more than one factor. Factorial designs are one way of examining these effects and assessing the extent of interactions. They also lend themselves to solution by the generalized least-squares matrix equation. The first-order model for a 2^2 factorial design without interaction is given by;

$$y_i = \beta_0 + \beta_1 x_{1i} + \beta_2 x_{2i} + r_i \qquad (5.12)$$

A matrix of parameter coefficients for this model has three columns:

$$X = \begin{bmatrix} x_0 & x_1 & x_2 \\ 1 & -1 & -1 \\ 1 & 1 & -1 \\ 1 & -1 & 1 \\ 1 & 1 & 1 \end{bmatrix}$$

in which the first column gives the parameter coefficients for the intercept, the second column gives the parameter coefficients for x_1, and the third column gives the parameter coefficients for x_2. This model assumes that the response surface is a plane in which there is no curvature or interaction.

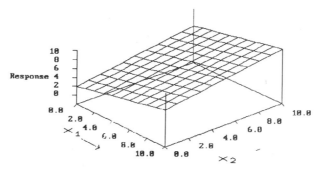

Fig. 5.5a. *Response surface for a two-factor model; $y = 2 + 0.4x_1 + 0.4x_2$*

Fig. 5.5a shows a typical three-dimensional first-order response surface for two factors described by the equation $y = 2 + 0.4x_1 + 0.4x_2$ plotted over the range 0 to 10 for both factors. These response surfaces are very attractive descriptions for systems in which two or more factors are important. Many software packages are capable of producing these (see Computer Software).

If interactions between the factors prove to be important the model can be changed to incorporate this and is then given by:

$$y_i = \beta_0 + \beta_1 x_{1i} + \beta_2 x_{2i} + \beta_{12} x_{1i} x_{2i} + r_i \qquad (5.13)$$

The matrix of parameter coefficients for this model becomes

$$X = \begin{bmatrix} x_0 & x_1 & x_2 & x_1x_2 \\ 1 & -1 & -1 & 1 \\ 1 & 1 & -1 & -1 \\ 1 & -1 & 1 & -1 \\ 1 & 1 & 1 & 1 \end{bmatrix}$$

in which the elements in the fourth column represent the parameter coefficients for the interaction between x_1 and x_2 (to give the estimate b_{12}).

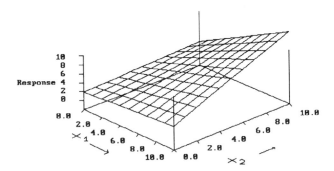

Fig. 5.5b. *Response surface for a two-factor model with interaction; y =*
$$2 + 0.1x_1 + 0.1x_2 + 0.1x_1x_2$$

Fig. 5.5b shows a three-dimensional response surface for two factors which interact as described by the equation $y = 2 + 0.1x_1 + 0.1x_2 + 0.1x_1x_2$. The response surface clearly tilts so that the slope taken by one of the factors depends on the value taken by the other factor.

The X^tX matrix for both the independent and the interacting model is given by:

$$X'X = \begin{bmatrix} 4 & 0 \\ 0 & 4 \end{bmatrix}$$

In Section 5.3, I stated that a special case for matrix inversion occurs where only the main diagonal elements are non-zero and that each element in the inverse is the reciprocal of the element in the original matrix. This is the case for all complete factorial designs and simplifies the manipulations involved. For example, the $X'X^{-1}$ matrix for the 2^2 factorial design is given by:

$$\left(X'X\right)^{-1} = \begin{bmatrix} 1/4 & 0 \\ 0 & 1/4 \end{bmatrix} = \frac{1}{4}I$$

where I is an identity matrix. The estimated parameter vector is now:

$$\hat{B} = \frac{1}{4}IX'Y = \frac{1}{4}X'Y \tag{5.14}$$

$$b_0 = \frac{1}{4}(+y_1 + y_2 + y_3 + y_4)$$

$$b_1 = \frac{1}{4}(-y_1 + y_2 - y_3 + y_4)$$

$$b_2 = \frac{1}{4}(-y_1 - y_2 + y_3 + y_4)$$

and

$$b_{12} = \frac{1}{4}(+y_1 - y_2 - y_3 + y_4)$$

This agrees with the way in which we calculated these effects in Part 3. There could also be a degree of curvature in the response surface for one or both of the factors. For example, Fig. 5.5c shows the three-dimensional response surface described by the equation $y = 0 + 0.01x_1 + 0.01x_2 + 0.05x_1^2 + 0.035x_2^2$.

The development of models to take account of possible curvature in three-dimensional response surfaces will be dealt with later on.

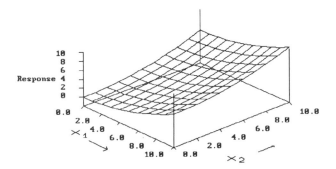

Fig 5.5c. *Response surface for a two-factor model with curvature; y =*
$$0 + 0.01x_1 + 0.01x_2 + 0.05x_1{}^2 + 0.035x_2{}^2$$

An important property of factorial designs is their orthogonality, in which there are no covariances among the estimated factor effects (b_1, b_2 etc.). There may be a degree of covariance between b_0 and any or all of the factor effects, but as long as there is no covariance among the factor effects, the design can be said to be orthogonal. A 2^2 design is therefore orthogonal because the off-diagonal elements of the $(X^tX)^{-1}$ matrix are all zero. In addition, all the parameters are free of confounding, the term applied when the parameters coefficients cannot be estimated separately.

5.5.1. Two-factor Example

In an investigation of the effects of two factors, burner height (x_1) and lamp current (x_2), upon the signal to noise ratio response of an atomic absorption spectrophotometer, a 2^2 factorial design with three additional points at the centre of the design was used. These centre points are often used in designs which require an estimate of purely experimental uncertainty. The conditions used were:

		Units	−1	0	+1
Burner height	x_1	(mm)	2	3	4
Lamp current	x_2	(mA)	3	4	5

The coded levels of the factors were found from Eq. (5.9) such that

$$x_{1i}* = (x_{1i} - 3)/1$$

and

$$x_{2i}* = (x_{2i} - 4)/1$$

The seven experimental runs were randomized and the responses obtained are given in Fig. 5.5d.

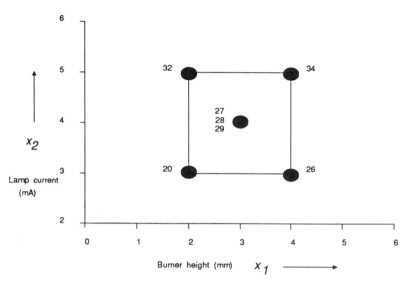

Fig. 5.5d. *Two-factor example design and the response obtained (signal to noise ratio)*

If we assume that the conditions used are well removed from the optimum, it is reasonable for us to adopt a first-order model for two factors, as given in Eq. (5.12). The purpose of this investigation is to optimize the signal to noise ratio, that is to maximize the response as a function of both factors. Therefore, we can imagine that we are at the bottom of a mountain and wish to reach the peak. The problem is we don't know the direction in which the mountain lies and determining this direction will be the initial aim of the investigation. At the bottom of this mountain it is unlikely that small changes in curvature will be very important and a first-order model may be adopted. The matrix of parameter coefficients for this model $(X*)$

and the vector of responses obtained (Y) are then as follows:

$$
X_* = \begin{bmatrix} x_0 & x_1 & x_2 \\ 1 & -1 & -1 \\ 1 & 1 & -1 \\ 1 & -1 & 1 \\ 1 & 1 & 1 \\ 1 & 0 & 0 \\ 1 & 0 & 0 \\ 1 & 0 & 0 \end{bmatrix} \qquad Y = \begin{bmatrix} 20 \\ 26 \\ 32 \\ 34 \\ 27 \\ 28 \\ 29 \end{bmatrix}
$$

We can calculate an estimate vector of parameters using the generalized least-squares matrix method (Eq. 5.4) as previously. The $X_*'X_*$ matrix is

$$
X_*'X_* = \begin{bmatrix} 1 & 1 & 1 & 1 & 1 & 1 & 1 \\ -1 & 1 & -1 & 1 & 0 & 0 & 0 \\ -1 & -1 & 1 & 1 & 0 & 0 & 0 \end{bmatrix} \begin{bmatrix} 1 & -1 & -1 \\ 1 & 1 & -1 \\ 1 & -1 & 1 \\ 1 & 1 & 1 \\ 1 & 0 & 0 \\ 1 & 0 & 0 \\ 1 & 0 & 0 \end{bmatrix}
$$

$$
= \begin{bmatrix} 7 & 0 & 0 \\ 0 & 4 & 0 \\ 0 & 0 & 4 \end{bmatrix}
$$

Again since only the main diagonal has non-zero elements, the inverse can be found by calculating the reciprocals of the individual elements:

$$
(X_*'X_*)^{-1} = \begin{bmatrix} 0.143 & 0 & 0 \\ 0 & 0.25 & 0 \\ 0 & 0 & 0.25 \end{bmatrix}
$$

This design is orthogonal because all the off-diagonal elements of this matrix are zero, making the corresponding elements in the variance–covariance matrix (V) zero. To estimate the parameters, we have to calculate the $X_*'Y$ matrix which is given by:

$$X *' Y = \begin{bmatrix} 1 & 1 & 1 & 1 & 1 & 1 & 1 \\ -1 & 1 & -1 & 1 & 0 & 0 & 0 \\ -1 & -1 & 1 & 1 & 0 & 0 & 0 \end{bmatrix} \begin{bmatrix} 20 \\ 26 \\ 32 \\ 34 \\ 27 \\ 28 \\ 29 \end{bmatrix} = \begin{bmatrix} 196 \\ 8 \\ 20 \end{bmatrix}$$

We can now estimate the parameters using Eq. (5.4):

$$\hat{B} = \begin{bmatrix} 0.143 & 0 & 0 \\ 0 & 0.25 & 0 \\ 0 & 0 & 0.25 \end{bmatrix} \begin{bmatrix} 196 \\ 8 \\ 20 \end{bmatrix} = \begin{bmatrix} 28 \\ 2 \\ 5 \end{bmatrix} \begin{aligned} &= b_0 \\ &= b_1 \\ &= b_2 \end{aligned}$$

The predicted least-squares regression equation for the coded system is therefore

$$y_i = 28 + 2x_{1i} * + 5x_{2i} * + r_i$$

These estimated parameters describe a plane with a slope of 2 units in the x_1* direction and a slope of 5 units in the x_2* direction with an estimated intercept (b_0) of 28 (ratio of signal to noise). This is shown as a 3-D response surface in Fig. 5.5e.

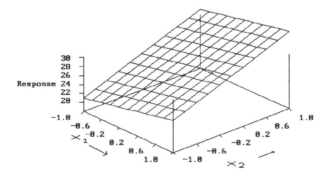

Fig. 5.5e. *Response surface for the two-factor model; $y_i = 28 + 2x_{1i} * + 5x_{2i}*$ plotted over the range -1 to $+1$ for both factors*

It is possible at this stage to carry out an analysis of variance on these results to examine whether the regression accounts for a significant proportion of the variance in the responses as I did for the single-factor designs, and also to test whether a lack of fit is significant at a particular probability level. However, since it was assumed that the centre of the design (0,0) was well removed from the optimum values for the factors, it is probably more useful to check this by examining whether first-order effects are much larger than second-order effects.

We can expand the model and recalculate the parameters, incorporating at the same time the coefficients to estimate curvature in the two directions as b_{11} and b_{22}, and an interaction between the two factors as b_{12}. However, the number of parameters in the model is then six. This uses up all the available degrees of freedom. Alternatively, we can check the extent of curvature and interaction using some simple procedures. In Part 3, it was shown that we can estimate interactions in factorial designs by applying the coefficients for the interaction to the responses. The first four experimental runs in the matrix of parameter coefficients compose a 2^2 factorial design, and these coefficients multiplied together are -1, 1, 1, and -1 respectively for the interaction. Applying these to the responses and dividing by 4 (number of runs in 2^2 factorial design) gives the interaction estimate:

$$b_{12} = (-20 + 26 + 32 - 34)/4 = 1$$

indicating a slight degree of interaction between these factors.

Curvature in both factors may be estimated jointly ($b_{11} + b_{22}$) if we imagine that the three points at the centre of the design are actually at the centre of a saucer-like surface. The four points composing the factorial design are then on the edge of this saucer. If there is no curvature in factors x_1* and x_2*, there should be little difference in the average responses of these two parts of the experimental design. Thus

$$b_{11} + b_{22} = \bar{y}_f - \bar{y}_c = \frac{(20 + 26 + 32 + 34)}{4} - \frac{(27 + 28 + 29)}{3} = 0$$

There is therefore no evidence of curvature in the responses.

So far, then, it has been shown that first-order effects predominate, with no curvature and only a slight interaction between burner height and lamp current. It is now possible to start to move towards the optimum. This can be accomplished by means of a technique known as *steepest ascent*.

5.6. STEEPEST ASCENT

The technique of steepest ascent is useful for moving towards the optimum in as few experimental runs as possible. Since curvature and interaction checks showed for this example that the first-order model is reasonable, it is possible for us to fit a planar response surface to the data in the region of factor space covered by the experimental design. Having done this, it should then be possible to move along this plane towards the optimum. Instead of a 3-dimensional diagram to represent the response surface, as in Fig. 5.5a, we can draw a contour diagram (rather like a contour map), with the contours representing equal responses (some software packages will only give this type of response surface representation). The co-ordinates for a particular contour are calculated from

$$y_0 = b_0 + b_1 x_1 * + b_2 x_2 * \qquad (5.15)$$

This equation can be used to find an x_1 co-ordinate when x_2 takes various values. Thus the response contour of $y = 28$ is given by

$$0 = 28 - 28 + 2x_1 * + 5x_2 *$$

For $x_1* = -1$ then $5x_2* = 2$, $x_2* = 2/5 = 0.4$. For $x_1* = 0$ then $5x_2* = 0$, $x_2* = 0$ and so on. These contours are straight lines which only require two points to be drawn in. Other response contour lines can be similarly calculated.

⊓ Calculate the response contour for signal to noise ratios of 30 and 32 by calculating x_2* values at x_1* values of -1, 0, and 1. Draw them in on Fig. 5.6a and work out the direction of the optimum.

The completed diagram (given at the end of the next sub-section) clearly shows the direction in which the experimental conditions need to move

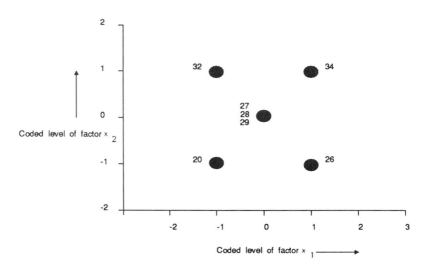

Incomplete Fig. 5.6a. *Contour diagram for first-order equation; y =* $28 + 2x_1 * + 5x_2*$ *showing direction of optimum*

in order to reach the optimum. It is at right angles to the contours. Since $b_1 = 2$ and $b_2 = 5$, the angle (θ) taken from the centre of the design may be calculated from

$$\tan \theta = b_2/b_1 = 5/2$$

$$\theta = 68.2 \text{ degrees}$$

Alternatively, the direction may be found by moving 5 units in the x_2 direction for every 2 units in the x_1 direction from the centre of the original design. Thus it is now possible to conduct an experiment at coded levels of $x_1* = 1$ and $x_2* = 2.5$. In this case it is relatively easy for us to convert the coded values for x_1 and x_2 back to their uncoded values. However, if this proves tricky because you have used intervals between the levels, you can manipulate Eq. (5.9) to yield the uncoded versions of the factors.

5.6.1. Stopping Rules

Experimental runs are usually continued along the path of steepest ascent until little progress is made or there is evidence that curvature should be taken into account. Given below are some general rules which you may use to decide whether any of these situations has arisen:

1. Responses remain static for two consecutive experiments.

2. Responses decrease for two consecutive experiments.

3. The differences between observed responses and those predicted from the least-squares regression equation (the residuals) are large.

This last rule is applied to detect curvature or a change in the direction of the plane of the underlying response surface. At this time, we must decide whether to carry out another first-order experiment in the region of the highest response, and perhaps move along a new path of steepest ascent, or to carry out a second-order design which can estimate the curvature and interaction.

If the interaction check for the first-order design used in the previous section had been more substantial and had been indicative of a underlying interaction, we might have followed a route not leading directly towards an optimum. Take a look, for example, at two response surfaces, one in which an interaction is absent (Fig. 5.6b) and one with an interaction present (Fig. 5.6c). Adopting a course which is perpendicular to the fitted contours

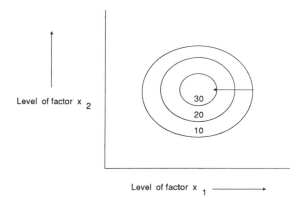

Fig. 5.6b. *Contour diagram of a two-factor response surface without interaction*

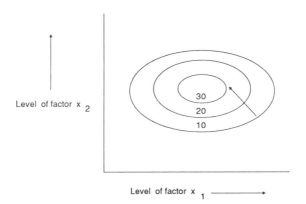

Fig. 5.6c. *Contour diagram of a two-factor response surface with interaction*

in the interaction case will not lead us directly to the optimum (Fig. 5.6c) whereas when the interaction is absent such a course will lead directly to the optimum (Fig. 5.6b).

Suppose we carry out the run at $x_1* = 1$ and $x_2* = 2.5$ as indicated above and this yields a signal to noise ratio response of 52, a distinct improvement on any response in the first-order design. Another run is now carried out at $x_1* = 2$ and $x_2* = 5$ which yields yet another improved S/N response of 83. However, when two further steps are taken along this apparently fruitful road at x_1*, x_2* co-ordinates of 3,7.5 and 4,10 respectively, the responses are not as high (*S/N* ratios of 53 and 42 respectively). Following the rules given above, we should now stop moving in this direction because the response is unlikely to increase beyond the largest value already achieved. Instead it is probably more reasonable to carry out a second-order experiment around $x_1* = 2$ and $x_2* = 5$.

5.6.2. Other Response Surface Techniques

Steepest ascent is not the only technique which can be used for moving towards an optimum, and one that is frequently used in chemical applications is *simplex optimization*. A simplex is a geometrical figure with $f + 1$ sides for f factors. Therefore, a simplex for two factors is a triangle

The points or vertices of the simplex are used as the co-ordinates for the experimental runs and the worst of the responses obtained is rejected. The position of a new vertex is calculated from a mirror image of the simplex from the position of the vertex which has been rejected. There are sets of rules which enable the experimenter to move towards the optimum. The simplex figure can move quite efficiently over the response surface towards an optimum, especially when a large number of factors are being considered. Unlike most classical response surface techniques the number of runs required by the simplex to move on and around the response surface is usually quite small once the initial $f + 1$ runs have been carried out. However, factorial and other statistically based response surface designs can be applied more readily when experiments are carried out in batches. The topic of simplex optimization is very interesting for many practising chemists but is too large to be considered in much detail here. The interested reader should try one or two of the references given in the bibliography.

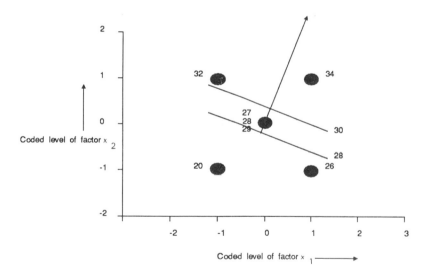

Completed Fig. 5.6a. *Contour diagram for first-order equation;* $y = 28 + 2x_1 * +5x_2*$ *showing direction of optimum*

5.7. SECOND-ORDER DESIGNS

The results for the steepest ascent procedure indicate that the optimum is somewhere near $x_1* = 2$, $x_2* = 5$, or in their uncoded versions at $x_1 = 5$

and $x_2 = 9$ respectively. It is likely that first-order effects are going to be somewhat less important in this region of factor space and a new design which can estimate the degrees of curvature and interactions more precisely is required. A suitable second-order model for such a case is

$$y_i = \beta_0 + \beta_1 x_{1i} + \beta_{11} x_{1i}^2 + \beta_2 x_{2i} + \beta_{22} x_{2i}^2 + \beta_{12} x_{1i} x_{2i} + r_i \qquad (5.16)$$

in which the second-order parameters to estimate curvature are β_{11} and β_{22} for factors x_1 and x_2 respectively, and β_{12} to represent the interaction. An experimental design constructed to estimate parameters for any approximating model should meet certain design criteria:

  (i) provide estimates for all the parameters required;

  (ii) require as few experimental runs as possible;

 (iii) provide a test for lack of fit;

  (iv) allow the experiment to be performed in blocks;

  (v) allow specified variance criteria for estimated parameters and estimated responses to be met.

Perhaps the most obvious class of design to estimate second-order effects for more than one factor is the 3^f factorial design, which satisfies several of the above design criteria. They are orthogonal designs but do require a relatively large number of runs even when fractional replicates are used.

5.7.1. Central Composite Designs

To overcome this problem, Box and Wilson added star designs to 2^f factorials to form central composite designs, the resulting design often being referred to as *Box–Wilson* designs. A typical central composite is shown in Fig. 5.7a.

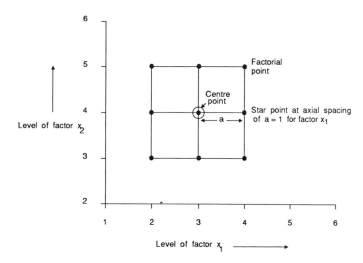

Fig. 5.7a. *Factor combinations for a central composite design for two factors with an axial spacing of a = ±1*

These designs are useful in that they do not require an excessive number of runs. They are based on two-level factorial designs or two-level fractional factorial designs (N_c runs) which have been augmented with N_o extra point(s) at the centre of the design and $2f = N_a$ (where f = number of factors) extra star points, one at either extreme of each factor and at the centre of all the other factors. Central composite designs based on complete factorial designs therefore require at least $2^f + 2f + 1$ runs. In practice, where an initial first-order design shows a significant lack of fit, it can be augmented with star points and extra centre points to form a central composite design.

∏   Is there any difference between a two-factor central composite design and a 3^2 factorial design?

In terms of the number of experimental runs required, this depends on the number of centre points used in the central composite design. At a minimum of one centre point there is no difference, since $3^2 = 9$ runs and $2^2 + 2 \times 2 + 1 = 9$ runs. However, there may also be some difference between the designs in terms of the spacing of the star points from the centre of the design. Fig. 5.7a illustrates at the same time a two-factor central composite with an axial spacing of ±1 in the star points, and a 3^2 factorial design. For the central composite design this axial spacing can vary, depending upon certain variance criteria. The minimum number of experimental runs required for a given number of factors by 3^f factorial designs and central composite designs is shown below:

Number of factors f	Treatment combinations	
	Three-level factorials 3^f	Composite $2^f + 2f + 1$
2	9	9
3	27	15
4	81	25
5	243	43
5	81 (1/3 fraction)	27 (1/2 fraction)
6	729	77
6	243 (1/3 fraction)	45 (1/2 fraction)

This table shows that the composite designs can achieve a great saving in the number of experimental runs required when the number of factors is large. For the two-factor example, however, they offer no advantage in this respect. They also suffer from the disadvantage that fewer degrees of freedom are available for estimating the residual error and that some of the parameters may be estimated with unequal variances. This has implications for the confidence we may place in the parameter estimates.

To construct a central composite design for a particular number of factors, values of N_o and "a" have to be determined. If we use a fractional factorial, it must be of resolution V (five) or greater to allow estimation of all the second-order coefficients. The number of centre points should be greater than one to estimate pure experimental uncertainty in a lack of fit test. How much greater than one is determined by the requirements for blocking, satisfying criteria for Var(b), and variance of the estimated responses. The value of "a" for axial points is determined by variance criteria for estimated parameters and responses. As with the first-order example, it is usual when working with central composite designs to code the lower and higher values of the points in the factorial design to -1 and $+1$ respectively. The centre of all the points then takes a value of zero for each of the factors.

As mentioned earlier the variance-covariance matrix, V, estimated by $(X*'X*)^{-1}s^2{}_{pe}$, has variances of the parameters down the main diagonal and the covariances between the parameters as off-diagonal elements. For a central composite design it is the values for N_c, N_a, N_o and the axial spacing (a) which determine this matrix. Careful selection of these values can lead to certain desirable properties for the design. A two-factor second-order model will be used to illustrate the effect upon the variance–covariance matrix by the specified coefficients for a central composite design. For this design the values chosen were $N_c = 4$, $N_a = 4$ with axial spacing of $a = 1.27$ and $N_o = 5$ centre points. The appropriate matrix of parameter coefficients is

$$X* = \begin{bmatrix} x_0 & x_1 & x_2 & x_1{}^2 & x_2{}^2 & x_1x_2 \\ 1 & +1 & +1 & +1 & +1 & +1 \\ 1 & +1 & -1 & +1 & +1 & -1 \\ 1 & -1 & +1 & +1 & +1 & -1 \\ 1 & -1 & -1 & +1 & +1 & +1 \\ 1 & +a & 0 & a^2 & 0 & 0 \\ 1 & -a & 0 & a^2 & 0 & 0 \\ 1 & 0 & +a & 0 & a^2 & 0 \\ 1 & 0 & -a & 0 & a^2 & 0 \\ 1 & 0 & 0 & 0 & 0 & 0 \\ 1 & 0 & 0 & 0 & 0 & 0 \\ 1 & 0 & 0 & 0 & 0 & 0 \\ 1 & 0 & 0 & 0 & 0 & 0 \\ 1 & 0 & 0 & 0 & 0 & 0 \end{bmatrix}$$

$$= \begin{bmatrix} x_0 & x_1 & x_2 & x_1{}^2 & x_2{}^2 & x_1x_2 \\ 1 & +1 & +1 & +1 & +1 & +1 \\ 1 & +1 & -1 & +1 & +1 & -1 \\ 1 & -1 & +1 & +1 & +1 & -1 \\ 1 & -1 & -1 & +1 & +1 & +1 \\ 1 & 1.27 & 0 & 1.61 & 0 & 0 \\ 1 & -1.27 & 0 & 1.61 & 0 & 0 \\ 1 & 0 & +1.27 & 0 & 1.61 & 0 \\ 1 & 0 & -1.27 & 0 & 1.61 & 0 \\ 1 & 0 & 0 & 0 & 0 & 0 \\ 1 & 0 & 0 & 0 & 0 & 0 \\ 1 & 0 & 0 & 0 & 0 & 0 \\ 1 & 0 & 0 & 0 & 0 & 0 \\ 1 & 0 & 0 & 0 & 0 & 0 \end{bmatrix} \quad (5.17)$$

The $X *^t X *$ matrix for a central composite design is formed in terms of $N = N_c + N_a + N_o$ and the axial spacing, and is given by:

$$X *^t X * = \begin{bmatrix} N & 0 & 0 & N_c + 2a^2 & N_c + 2a^2 & 0 \\ 0 & N_c + 2a^2 & 0 & 0 & 0 & 0 \\ 0 & 0 & N_c + 2a^2 & 0 & 0 & 0 \\ N_c + 2a^2 & 0 & 0 & N_c + 2a^4 & N_c & 0 \\ N_c + 2a^2 & 0 & 0 & Nc & N_c + 2a^4 & 0 \\ 0 & 0 & 0 & 0 & 0 & N_c \end{bmatrix} \quad (5.18)$$

The $X *^t X *$ matrix for the two-factor example is therefore

$$X *^t X * = \begin{bmatrix} 13 & 0 & 0 & 7.21 & 7.21 & 0 \\ 0 & 7.21 & 0 & 0 & 0 & 0 \\ 0 & 0 & 7.21 & 0 & 0 & 0 \\ 7.21 & 0 & 0 & 9.154 & 4 & 0 \\ 7.21 & 0 & 0 & 4 & 9.154 & 0 \\ 0 & 0 & 0 & 0 & 0 & 4 \end{bmatrix}$$

Unfortunately, we cannot find the inverse of this matrix using any of the previously described methods because of the increased dimensions. Inverses of matrices which are larger than 3×3 have to be calculated either with a computer program or by a rather complex set of algebraic equations. For the example given above with $N_c = 4$ factorial points, $N_a = 4$ star points at an axial spacing of ± 1.27 and $N_o = 5$ centre points, this is given by:

$$(X *^t X *)^{-1} = \begin{bmatrix} 0.196 & 0 & 0 & -0.108 & -0.108 & 0 \\ 0 & 0.139 & 0 & 0 & 0 & 0 \\ 0 & 0 & 0.139 & 0 & 0 & 0 \\ -0.108 & 0 & 0 & 0.198 & 0.042 & 0 \\ -0.108 & 0 & 0 & 0.042 & 0.198 & 0 \\ 0 & 0 & 0 & 0 & 0 & 0.25 \end{bmatrix}$$

From the $(X *^t X *)^{-1}$ matrix it can be seen that the covariances between the estimated intercept and first-order parameters and the estimated second-order mixed parameters will be zero whereas covariances between estimated intercept and estimated pure second-order parameters, $Cov(b_0, b_{jj})$, will not be zero. Covariances between estimates of pure second-order parameters, $Cov(b_{11}, b_{22})$, will be zero, so giving an orthogonal design, if it is set up so that

$$NN_c = \left[N_c + 2a^2 \right]^2 \quad (5.19)$$

This condition for uncorrelated, estimated, pure second-order parameters is true for any number of factors.

A central composite design for three factors is given in Fig. 5.7b and clearly shows the factorial, star and centre-point components.

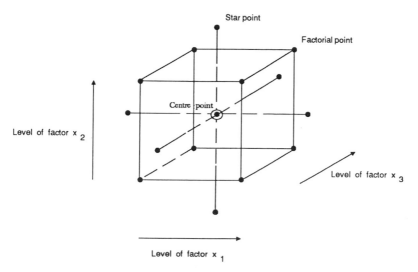

Fig. 5.7b. *Factor combinations for a central composite design in three dimensional factor space*

It is possible to calculate the $(X_*'X_*)^{-1}$ matrix of the central composite for any number of factors but this requires an increase in the dimensions of the $X_*'X_*$ matrix to incorporate more factors and therefore more parameters, making it even more difficult to calculate manually. It is therefore only feasible to calculate these inverses using computer programs.

The axial spacing needed to ensure orthogonality may be calculated from

$$a^2 = \frac{\sqrt{(N_c + N_a + N_o)N_c} - N_c}{2} \tag{5.20}$$

For example, to obtain an orthogonal design for two factors with $N_c = 4$, $N_a = 4$ and $N_o = 1$, the axial spacing should be $a = \pm 1$. This design is equivalent to the 3^2 factorial design. If the number of centre points is increased from 1 to 5 then the axial spacing has to be increased to $a = \pm 1.27$, as for the above two-factor composite design.

∏   If the number of centre points in the two-factor central composite
       design is increased from $N_o = 1$ to $N_o = 2$, what is the axial spacing
       now required to maintain orthogonality?

Using the above formula,

$$a^2 = \frac{\sqrt{(4 + 4 + 2)4} - 4}{2} = \frac{\sqrt{40} - 4}{2} = 1.162$$

$$a = \sqrt{1.162} = 1.078$$

Therefore, the axial spacing would have to be increased to 1.078 to maintain
orthogonality. Orthogonal designs are almost always produced by coding the
factor values to ± 1 and have the property that the mathematics involved is
very simple, as in the case of the factorial designs. However, transformation
of the estimated parameters back to the original uncoded factor space
usually destroys orthogonality.

Orthogonality eliminates covariances between the individual estimated
pure second-order parameters. However, rather than using a criterion for
individual parameters, it is possible to use criteria based on the joint effect
of all the parameters. One such criterion is based on variances of estimated
responses for points that are an equal distance from the design centre.
Designs that have points which are equidistant from the centre are termed
rotatable designs. This criterion means that the variance of the predicted
response depends only on the distance from the design centre and not on
the direction. A necessary condition for a design to be rotatable is that the
axial spacing for a central composite design is the fourth root of the number
of cube points given by

$$a^4 = N_c \tag{5.21}$$

or

$$a^2 = \sqrt{N_c}$$

Therefore, rotatability does not depend on the number of centre points. We
can add centre points to orthogonal designs with a corresponding increase
in the axial spacing to satisfy both rotatability and orthogonality.

Given below are the numbers of points required in each component of central composite designs which satisfy rotatability and approach orthogonality. For every value of f factors the number of points in the factorial and star portions of the design is defined as previously.

Rotatable and orthogonal designs for f factors:

f	N_c	N_a	N_o	Axial spacing
2	4	4	8	1.414
3	8	6	9.3	1.682
4	16	8	12	2.000
5	32	10	16.6	2.378
6	64	12	24	2.828

Clearly, rotatability and orthogonality cannot always be achieved in the same central composite design. One of the easiest ways for us to design experiments which are both orthogonal and rotatable is to work out the axial spacing to satisfy rotatability and then add as many centre points as possible to approach orthogonality.

SAQ 5.7a	What is the axial spacing required for a two-factor central-composite rotatable design with eight centre points? Is this design orthogonal?

SAQ 5.7a

5.7.2. Running Central Composite Designs in Blocks

If we have insufficient raw material or time available to ensure that all the experimental units or runs can be carried out under uniform conditions, we may have to divide the design into blocks. The effect of blocking is assumed to introduce a constant value to the response in a particular block, and the designs have to be set up so that additive block effects are not confounded with coefficients of the second-order model. Getting rid of this confounding by blocking is termed *orthogonal blocking*. It implies that the columns of coefficients for blocking are not confounded with any of the columns for the first- and second-order coefficients, a condition which means that each block must be a first-order orthogonal block.

Central composite designs are usually divided so that cube points are in one block and axial points in another block. Any further blocking requirements are now met by partitioning the cube points into fractional factorials of resolution III (three). We have to partition the centre points between the cube block (N_{oc}) and the axial block (N_{oa}) and we may have to take decisions as to how many and in what ratio they should be allocated. Once we have decided this we have to calculate the axial spacing to maintain orthogonality.

$$a^2 = \frac{N_c(2f + N_{oa})}{2(N_c + N_{oc})} \qquad (5.22)$$

For a central composite design constructed for two factors in two blocks with four centre points divided equally between the two blocks, $N_c = 4$, $N_a = 4$, $N_{oa} = 2$ and $N_{oc} = 2$. For orthogonal blocking

$$a^2 = 4(4 + 2)/[2(4 + 2)] = 2$$

and therefore

$$a = \pm 1.414$$

We then run the design in two blocks, one block with the factorial design points and two centre points, the other block with two centre points and the axial points at an axial spacing of $a = \pm 1.414$. We can also divide the centre points so that other design criteria, such as rotatability, are satisfied, making it possible to partition the centre points and then work out the axial spacing required for rotatability.

5.7.3. Worked Example Central Composite Design

Continuing the investigation of the effects of burner height (x_1) and lamp current (x_2) upon the signal to noise ratio response of an atomic absorption spectrophotometer, it was decided to use a second-order model (Eq. 5.16) and an orthogonal two-factor central composite design with five centre points and an axial spacing of ± 1.27. This required a total of 13 experimental runs.

The result of following the path of steepest ascent was that the optimum was thought to occur around $x_1 = 5$ mm and $x_2 = 9$ mA. These co-ordinates were used as the new centre point for the design. An interval of ± 1 was retained for both factors. Therefore, the minimum and maximum values of factors x_1 and x_2 were:

	Coded values	−1	0	+1
Burner height	x_1 (mm)	4	5	6
Lamp current	x_2 (mA)	8	9	10

The appropriate matrix of parameter coefficients is as given in Eq. (5.17) and the responses obtained for the runs (in the order in which they were carried out) are given in Fig. 5.7c.

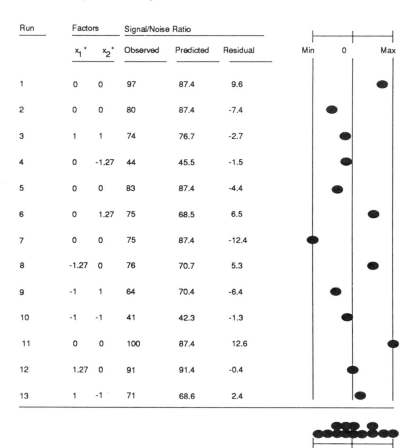

Run	Factors		Signal/Noise Ratio					
	x_1^*	x_2^*	Observed	Predicted	Residual	Min	0	Max
1	0	0	97	87.4	9.6			
2	0	0	80	87.4	-7.4			
3	1	1	74	76.7	-2.7			
4	0	-1.27	44	45.5	-1.5			
5	0	0	83	87.4	-4.4			
6	0	1.27	75	68.5	6.5			
7	0	0	75	87.4	-12.4			
8	-1.27	0	76	70.7	5.3			
9	-1	1	64	70.4	-6.4			
10	-1	-1	41	42.3	-1.3			
11	0	0	100	87.4	12.6			
12	1.27	0	91	91.4	-0.4			
13	1	-1	71	68.6	2.4			

Fig. 5.7c. *Design matrix, observed responses and responses predicted from the second-order equation, and residuals plotted as a frequency distribution*

The estimated parameters were determined by the generalized least-squares matrix solution (Eq. 5.4). These parameters give the second-order equation describing the response as a function of the coded factor levels:

$$y_i = 87.39 + 8.18x_{1i} * +9.05x_{2i} * -3.98x_{1i} *^2 -18.98x_{2i} *^2 -4.97x_{1i} * x_{2i}*$$

Predicted values of the S/N ratios were calculated and subtracted from the observed S/N ratios to give the residuals in Fig. 5.7c. Reassuringly, these residuals seem to have occurred randomly and when plotted as a frequency distribution are approximately normally distributed.

Fig. 5.7d is a three-dimensional response surface for the above equation plotted over the -1 to $+1$ values for both factors used in the coded system. The values taken by the parameter estimates dictate the shape of the response surface. The signs of the pure second-order terms in this equation indicate that there is a stationary point on the response surface which represents a maximum for the two factors. This may be used as the optimum combination of burner height and lamp current. The equation also indicates that the estimated response surface will fold quadratically downwards away from this stationary point.

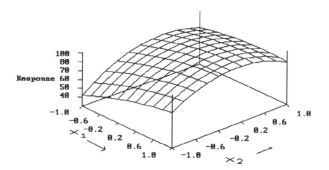

Fig. 5.7d. *Response surface for the two-factor example;* $y_i = 87.39 + 8.18x_{1i} * +9.05x_{2i} * -3.98x_{1i} *^2 -18.98x_{2i} *^2 -4.97x_{1i} * x_{2i}*$ *plotted over the coded levels -1 to $+1$.*

An analysis of variance table has been constructed to examine whether the model accounts for a significant proportion of the total variance in the data:

Source of variation	d.f.	SS	MS	Variance ratio
Total (corrected)	12	3708.77		
Regression	5	3101.15	620.23	7.15
Residual error	7	607.62	86.80	
Lack of fit	3	128.85	42.95	0.36
Pure error	4	478.77	119.69	

The sum of squares due to the regression as a percentage of the total sum of squares is 83.6%, showing that a reasonably large proportion of the variance is explained by the regression equation. The variance ratio of the regression mean square to the residual mean square gives a value of 7.15, which is significant at a probability level of $P = 0.05$, indicating that the regression equation accounts for a reasonable proportion of the variance in the responses.

As indicated earlier, residual error in response surface experiments can be broken down into two components, a lack of fit and pure experimental uncertainty. In this case, the *PureSS* has been calculated from the responses at the five centre points. The *LofSS* has been calculated by subtracting the *PureSS* from the *ResidSS*. The ratio of these mean squares is 0.36, a low value which indicates that the second-order model is an adequate approximation to the data. The residual mean square of 86.80 can thus reasonably be used as an estimate of the variance due to pure experimental uncertainty (s^2_{pe}) and hence used to calculate the variances and covariances between the estimated parameters as shown earlier. The values on the principal diagonal of the $(X_*'X_*)^{-1}$ matrix have therefore been multiplied by s^2_{pe} to yield the variances.

$$
\begin{aligned}
Var(b_0*) &= (0.196)(119.69) = 23.46 \\
Var(b_1*) &= (0.139)(119.69) = 16.64 \\
Var(b_2*) &= (0.139)(119.69) = 16.64 \\
Var(b_{11}*) &= (0.198)(119.69) = 23.70 \\
Var(b_{22}*) &= (0.198)(119.69) = 23.70 \\
Var(b_{12}*) &= (0.25)\ (119.69) = 29.92
\end{aligned}
$$

The square roots of these variances provide standard errors of the parameter estimates which can now be individually tested by means of t-tests to see if they are significantly different from zero. This is accomplished as indicated earlier by calculating ratios of the estimated parameters to their standard errors to give t-values which may be compared with tabulated values of t at different probability levels. These probability values are given below:

Parameter	b_0*	b_1*	b_2*	$b_{11}*$	$b_{22}*$	$b_{12}*$
Estimate	87.39	8.18	9.05	−3.98	−18.98	−4.97
Standard error	4.84	4.08	4.08	4.87	4.87	5.47
t-value	18.06	2.00	2.22	−0.97	−3.90	−0.91
%Probability <> 0	>99.5	>90	>95	<90	>99.5	<90

From the probability values, most of the estimated parameters would seem to be non-zero, confirming that the model is an adequate fit. Of course, where an overall test of fit for the regression equation confirms that the model is adequate, it is not very important to test whether that these individual parameter estimates are significantly different from zero or not. However, it may be useful when the predicted regression equation fits the data poorly so that a new model can be proposed.

The uncoded values of the parameter estimates may be calculated using matrix equations (5.10) and (5.11):

$$
\widehat{B} = T^{-1}\widehat{B}*
$$

in which T is a transform matrix given by

$$T = (X *^t X*)^{-1}(X *^t X)$$

The matrix $(X*'X*)^{-1}$ was calculated earlier and the matrix $X*'X$ is calculated below:

$$X *^t X = \begin{bmatrix} 1 & 1 & 1 & 1 & 1 & 1 & 1 & 1 & 1 & 1 & 1 & 1 & 1 \\ 1 & 1 & -1 & -1 & 1.27 & -1.27 & 0 & 0 & 0 & 0 & 0 & 0 & 0 \\ 1 & -1 & 1 & -1 & 0 & 0 & -1.27 & 1.27 & 0 & 0 & 0 & 0 & 0 \\ 1 & 1 & 1 & 1 & 1.61 & 1.61 & 0 & 0 & 0 & 0 & 0 & 0 & 0 \\ 1 & 1 & 1 & 1 & 0 & 0 & 1.61 & 1.61 & 0 & 0 & 0 & 0 & 0 \\ 1 & -1 & -1 & 1 & 0 & 0 & 0 & 0 & 0 & 0 & 0 & 0 & 0 \end{bmatrix}$$

$$\begin{bmatrix} 1 & 6 & 10 & 36 & 100 & 60 \\ 1 & 6 & 8 & 36 & 64 & 48 \\ 1 & 4 & 10 & 16 & 100 & 40 \\ 1 & 4 & 8 & 16 & 64 & 32 \\ 1 & 6.27 & 9 & 39.3 & 81 & 56 \\ 1 & 3.73 & 9 & 13.9 & 81 & 34 \\ 1 & 5 & 10.27 & 25 & 105 & 51 \\ 1 & 5 & 7.73 & 25 & 60 & 39 \\ 1 & 5 & 9 & 25 & 81 & 45 \\ 1 & 5 & 9 & 25 & 81 & 45 \\ 1 & 5 & 9 & 25 & 81 & 45 \\ 1 & 5 & 9 & 25 & 81 & 45 \\ 1 & 5 & 9 & 25 & 81 & 45 \end{bmatrix}$$

Therefore,

$$X *^t X = \begin{bmatrix} 13.00 & 65.00 & 117.00 & 332.21 & 1060.20 & 585.00 \\ 0.00 & 7.21 & 0.00 & 72.16 & 0.00 & 64.88 \\ 0.00 & 0.00 & 7.21 & 0.00 & 129.77 & 36.10 \\ 7.22 & 36.10 & 64.98 & 189.66 & 588.82 & 324.90 \\ 7.22 & 36.10 & 64.98 & 184.50 & 593.97 & 324.90 \\ 0.00 & 0.00 & 0.00 & 0.00 & 0.00 & 4.00 \end{bmatrix}$$

and

$$T = (X *' X*)^{-1} X *' X = \begin{bmatrix} 1.00 & 5.00 & 9.00 & 25.00 & 81.00 & 45.00 \\ 0.00 & 1.00 & 0.00 & 10.01 & 0.02 & 9.00 \\ 0.00 & 0.00 & 1.00 & 0.00 & 18.00 & 5.01 \\ 0.00 & 0.00 & 0.00 & 1.00 & 0.00 & 0.00 \\ 0.00 & 0.00 & 0.00 & 0.00 & 1.00 & 0.00 \\ 0.00 & 0.00 & 0.00 & 0.00 & 0.00 & 1.00 \end{bmatrix}$$

The inverse T^{-1} is best calculated by a suitable computer program. This then premultiplies the estimated matrix of coded parameters to give the uncoded versions of the parameter estimates.

$$\hat{B} = \begin{bmatrix} 1.00 & -5.00 & -9.00 & 25.01 & 81.26 & 45.04 \\ 0.00 & 1.00 & 0.00 & -10.03 & 0.02 & -8.99 \\ 0.00 & 0.00 & 1.00 & 0.02 & -18.07 & -5.01 \\ 0.00 & 0.00 & 0.00 & 1.00 & 0.00 & 0.00 \\ 0.00 & 0.00 & 0.00 & 0.00 & 1.00 & 0.00 \\ 0.00 & 0.00 & 0.00 & 0.00 & 0.00 & 1.00 \end{bmatrix} \begin{bmatrix} 87.39 \\ 8.18 \\ 9.50 \\ -3.98 \\ -18.98 \\ -4.97 \end{bmatrix}$$

$$= \begin{bmatrix} -1895.25 \\ 92.70 \\ 375.49 \\ -3.98 \\ -18.98 \\ -4.97 \end{bmatrix}$$

The equation describing a response in uncoded factor space is now given by:

$$y_i = -1895.25 + 92.70x_{1i} + 375.49x_{2i} - 3.98x_{1i}^2 - 18.98x_{2i}^2 - 4.97x_{1i}x_{2i}$$

It is possible to plot the response surface for this equation in the same way as has been done for other designs. However, the example central composite design used had the very desirable property of orthogonality, and conversion of the coded factor space to uncoded factor space destroys this property. Nevertheless, you should note at this point that the values taken by the second-order terms are exactly the same as in the coded equation given earlier.

5.7.4. Optimization

Having produced a second-order equation which describes the response as a function of the two factors, it is possible to pick out certain types of response surface without plotting a contour map or a 3-D response surface by examining the pure second-order terms. These should tell you the approximate shape of the response surface and whether or not an optimization of the factors is possible. If both b_{11} and b_{22} are positive and of approximately the same magnitude, the response surface will be a parabola opening upwards as shown in Fig. 5.7e.

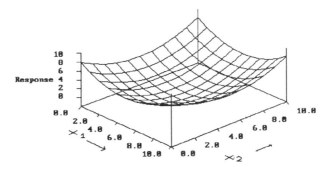

Fig. 5.7e. *3-D response surface with parabola opening upwards*

The stationary point in this situation represents a minimum on this response surface and is therefore not very interesting if the aim is to optimize the factors. If both b_{11} and b_{22} are negative and of approximately the same magnitude, the response surface will be a parabola opening downwards, in which case the stationary point represents a maximum, as in the atomic absorption example. This is shown in Fig. 5.7f.

Where both b_{11} and b_{22} are negative with a slight difference in their magnitudes, the response surface becomes flatter. However, if they are both negative but very different a ridged response surface is likely, as shown in Fig. 5.7g.

When one of the parameters is negative and the other positive and both are about the same size, a saddle is being described (Fig. 5.7h).

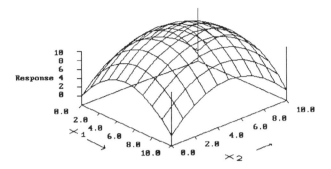

Fig. 5.7f. *3-D response surface with parabola opening downwards*

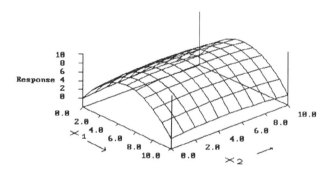

Fig. 5.7g. *3-D response surface with ridge*

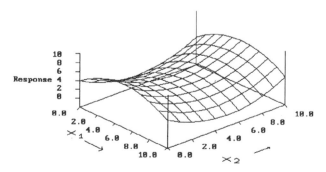

Fig. 5.7h. *3-D response surface with saddle*

To obtain the co-ordinates of the stationary point on the response surface we have to carry out a *canonical analysis* of the second-order equations derived from the models discussed earlier. The second-order equations are reduced to canonical forms by the technique of analytical geometry which moves the origin of the factor space to the stationary point on the estimated response surface, the point at which the partial derivatives of the response are set to zero for each factor. This gets rid of the first-order terms. Next, the factor axes are rotated to remove the interaction terms so that new axes coincide with the main axes of the estimated response surface. The polynomial equation is then differentiated with respect to each of the factors and the derivatives are set to zero. For the two-factor example this yields two equations:

$$\delta y / \delta x_1 = b_1 + 2b_{11}x_1 + b_{12}x_2 = 0 \qquad (5.23)$$

$$\delta y / \delta x_2 = b_2 + 2b_{22}x_2 + b_{12}x_1 = 0$$

At the stationary point two equations that satisfy this condition are given by;

$$2b_{11}x_{1s} + b_{12}x_{2s} = -b_1 \qquad (5.24)$$

$$b_{12}x_{1s} + 2b_{22}x_{2s} = -b_2$$

in which x_{1s} and x_{2s} are the co-ordinates of the stationary point. These equations may be also be expressed in matrix notation:

$$\begin{bmatrix} 2b_{11} & b_{12} \\ b_{12} & 2b_{22} \end{bmatrix} \begin{bmatrix} x_{1s} \\ x_{2s} \end{bmatrix} = \begin{bmatrix} -b_1 \\ -b_2 \end{bmatrix} \qquad (5.25)$$

The first matrix is one containing only second-order parameters, and dividing this by a scalar 2 yields a matrix S of $f \times f$ dimensions in which the single-factor second-order parameter estimates lie along the main diagonal. For the two-factor central composite design example:

$$S = \begin{bmatrix} b_{11} & b_{12}/2 \\ b_{12}/2 & 2b_{22} \end{bmatrix} = \begin{bmatrix} -3.98 & -2.49 \\ -2.49 & -18.98 \end{bmatrix}$$

The column vector of the stationary point co-ordinates may be defined as s:

$$s = \begin{bmatrix} x_{1s} \\ x_{2s} \end{bmatrix}$$

and the column vector of the first-order parameter estimates is defined as f:

$$f = \begin{bmatrix} b_1 \\ b_2 \end{bmatrix}$$

The matrix equation is therefore reduced to

$$2Ss = -f \quad \text{or} \quad Ss = -0.5f \tag{5.26}$$

which may be rearranged to yield the stationary point co-ordinates:

$$s = -0.5S^{-1}f \tag{5.27}$$

The stationary point co-ordinates may now be found using this equation:

$$s = \begin{bmatrix} x_{1s} \\ x_{2s} \end{bmatrix} = -0.5 \begin{bmatrix} -3.98 & -2.49 \\ -2.49 & -18.98 \end{bmatrix}^{-1} \begin{bmatrix} 92.70 \\ 375.49 \end{bmatrix}$$

The determinant of S is given by $(-3.98 \times -18.98) - (-2.49 \times -2.49) = 69.3$. This enables us to calculate the inverse matrix S^{-1} and therefore the stationary point co-ordinates.

$$s = \begin{bmatrix} x_{1s} \\ x_{2s} \end{bmatrix} = -0.5 \begin{bmatrix} -18.98/69.3 & 2.49/69.3 \\ 2.49/69.3 & -3.98/69.3 \end{bmatrix} \begin{bmatrix} 92.70 \\ 375.49 \end{bmatrix}$$

$$= -0.5 \begin{bmatrix} -0.274 & 0.036 \\ 0.036 & -0.057 \end{bmatrix} \begin{bmatrix} 92.70 \\ 375.49 \end{bmatrix} = \begin{bmatrix} 5.95 \\ 9.03 \end{bmatrix}$$

The stationary point of the response surface is to be found at $x_{1s} = 5.95$ and $x_{2s} = 9.03$. We may calculate the response at this stationary point by substituting these co-ordinates into the uncoded equation describing the response surface. This was found to be 91.45.

To produce three-dimensional response surface representations of equations with more than three factors would require four dimensions, one for each of the factors and a fourth for the response. Clearly this is not possible.

However, it is possible to have 3-D diagrams for two of the factors at a time, keeping the third constant. Once the stationary point has been calculated for any factor, it is this value which can be used in the equation. Thus three 3-D response surface diagrams can be produced for a three-factor model.

Translation of the origin of factor space to the stationary point on the response surface has the effect of making the first-order effects equal zero, with the estimated second-order parameters b_{11}, b_{22} and b_{12} for the two-factor model having the same values as in the uncoded equation. Rotation to make the mixed second-order terms in matrix S equal zero is an eigenvector and eigenvalue problem which will not be dealt with here because of its complexity. Suffice it to say that the rotated factor axes, known as canonical factors, describe the main directions of the second-order response surface and can reveal the essential features of the response surface and therefore allow a simpler interpretation of the effects of the factors.

SAQ 5.7b	What are the co-ordinates of the stationary point on the response surface for the equation given below? $$y_i = 5 + 2x_{1i} + 2x_{2i} + 3x_{1i}^2 + 3x_{2i}^2 - 4x_{1i}x_{2i}$$ What is the response at this stationary point? What shape will the response surface take?

Summary

I hope you now understand some of the basic principles behind response surface experimentation and exploration. This was not intended to be a completely comprehensive coverage of the topic, but you should at least have some ideas about models for single-factor and multi-factor designs, how to analyse them and interpret the results in a meaningful way. You should be then able to decide upon the best design to answer your most searching questions. Please also remember that although many of the techniques used in this and other parts of the book can be carried out automatically by suitable computer packages, it is necessary that you understand the principles behind the designs and calculations involved and do not accept output from them at face value.

Learning Objectives

After reading the material in the latter sections of Part 5 you should now be able to:

- understand how first-order and second-order models may be developed when more than one factor is to be examined;

- calculate estimates of parameters for two-factor models using the generalized least-squares matrix solution;

- realize that there are limitations to factorial designs as tools for multi-factor response-surface investigations;

- set up, analyse and interpret the results of central composite designs;

- interpret the response surface from the estimated parameters in a second-order equation for two factors;

- calculate the co-ordinates of a stationary point on the response surface and its corresponding response.

SAQs AND RESPONSES FOR PART FIVE

SAQ 5.3a

The following statements refer to different models and hypotheses. Indicate whether the statements are true (T) or false (F).

(i) In the model $y_i = \beta_1 x_{1i} + r_i$ there will be no intercept and the data will be forced through zero.   [True/False]

(ii) When the model fits the data perfectly, estimates of the predicted responses given by the matrix equation

$$\widehat{Y} = X\widehat{B}$$

will reproduce the original responses.
[True/False]

(iii) In the model $y_i = 0 + r_i$, the experimenter wishes to test the hypothesis that the intercept is significantly different from zero.[True/False]

(iv) When the model $y_i = \beta_0 + \beta_1 x_{1i}$ is used for a number of experimental runs all at the same level of x_1, the intercept can be zero. [True/False]

(v) The experimenter can be at least 95% confident in rejecting the null hypothesis if the calculated value of F (variance ratio) exceeds the tabulated value at the $P = 0.05$ probability level.   [True/False]

Response

(i) T. β_1, the slope in the direction x_1, is the only parameter in this model. Any least-squares line calculated on the basis of this model will be forced through zero with the possible consequence that the residuals will be inflated.

(ii) T. In the case of a perfect fit, the model for the matrix equation

$$Y = X\widehat{B} + R$$

would become

$$Y = X\widehat{B}$$

The estimates of the parameters are given by

$$\widehat{B} = X^{-1}Y$$

The relationship $\widehat{Y} = X\widehat{B}$ then reproduces the original responses exactly since the residuals are all zero.

(iii) F. The model contains no parameters. Therefore, parameters such as an intercept or slope cannot enter any hypothesis. The model attempts to establish whether the responses obtained are different from zero.

(iv) F. In this case the only slope which is possible is one of infinity, with no intercept because all the responses must vary along a line passing through the only level of x_1 used.

(v) T. Because it is quite likely that the calculated value corresponds to a greater than 95% probability value.

****

SAQ 5.3b

In an experiment to investigate the effect of pH upon the concentration of a complex species, the pH levels used were 2.0, 3.0 and 4.0, and the observed concentrations of a species Y were 0.30, 0.46 and 0.59 (mol dm^{-3}). Determine the parameters for a first-order model incorporating the slope and intercept, and decide by means of an analysis of variance whether the model can be accepted as a reasonable fit of the data at a probability level of $P = 0.05$.

Response

Given that three pH levels were investigated, the design matrix to be used for a first-order model would incorporate an intercept (β_0) and a first-order slope (β_1). The pH levels were 2, 3 and 4, so yielding the design matrix D and the matrix of parameter coefficients X given below:

$$D = \begin{bmatrix} 2 \\ 3 \\ 4 \end{bmatrix} \qquad X = \begin{bmatrix} x_0 & x_1 \\ 1 & 2 \\ 1 & 3 \\ 1 & 4 \end{bmatrix}$$

The $X^t X$ matrix is therefore

$$X^t X = \begin{bmatrix} 1 & 1 & 1 \\ 2 & 3 & 4 \end{bmatrix} \begin{bmatrix} 1 & 2 \\ 1 & 3 \\ 1 & 4 \end{bmatrix} = \begin{bmatrix} 3 & 9 \\ 9 & 29 \end{bmatrix}$$

The $(X^t X)^{-1}$ inverse is calculated using the determinant $(3 \times 29 - 9 \times 9) = 87 - 81 = 6$. Applying this determinant to the rearranged elements of the $X^t X$ matrix gives the inverse

$$\left(X^t X\right)^{-1} = \begin{bmatrix} 29/6 & -9/6 \\ -9/6 & 3/6 \end{bmatrix} = \begin{bmatrix} 4.833 & -1.5 \\ -1.5 & 0.5 \end{bmatrix}$$

Also,

$$X^t Y = \begin{bmatrix} 1 & 1 & 1 \\ 2 & 3 & 4 \end{bmatrix} \begin{bmatrix} 0.30 \\ 0.46 \\ 0.59 \end{bmatrix} = \begin{bmatrix} 1.35 \\ 4.34 \end{bmatrix}$$

The generalized least-squares solution (Eq. 5.4) for this example is then

$$\hat{B} = \left(X'X\right)^{-1} \qquad X'Y$$

$$\hat{B} = \begin{bmatrix} 4.833 & -1.5 \\ -1.5 & 0.5 \end{bmatrix} \begin{bmatrix} 1.35 \\ 4.34 \end{bmatrix} = \begin{bmatrix} 0.015 \\ 0.145 \end{bmatrix} \begin{matrix} = b_0 \\ = b_1 \end{matrix}$$

Therefore, any response may be represented by the equation

$$y_i = 0.015 + 0.145x_{1i} + r_i$$

If you didn't have these values, first of all recheck your calculations. If you then still have problems re-do Section 5.3.1 and pay particular attention to the matrix manipulations.

These parameter estimates can be used to obtain the predicted responses at the levels used in the experiment from

$$\hat{Y} = X\hat{B}$$

Therefore,

$$\hat{Y} = \begin{bmatrix} 1 & 2 \\ 1 & 3 \\ 1 & 4 \end{bmatrix} \begin{bmatrix} 0.015 \\ 0.145 \end{bmatrix} = \begin{bmatrix} (1 \times 0.015) + (2 \times 0.145) \\ (1 \times 0.015) + (3 \times 0.145) \\ (1 \times 0.015) + (4 \times 0.145) \end{bmatrix} = \begin{bmatrix} 0.305 \\ 0.45 \\ 0.595 \end{bmatrix}$$

Subtracting $\hat{Y}$ from Y gives the residuals R, which can be used in an analysis of variance to test the adequacy of the first-order model for the observed data,

$$Y \quad - \quad \hat{Y} \quad = \quad R$$

$$\begin{bmatrix} 0.30 \\ 0.46 \\ 0.59 \end{bmatrix} - \begin{bmatrix} 0.305 \\ 0.45 \\ 0.595 \end{bmatrix} = \begin{bmatrix} -0.005 \\ 0.01 \\ -0.005 \end{bmatrix}$$

The values in the R matrix can be squared and summed to give the sum of squares of the residuals (*ResidSS*).

$$R'R = \begin{bmatrix} -0.005 & 0.01 & -0.005 \end{bmatrix} \begin{bmatrix} -0.005 \\ 0.01 \\ -0.005 \end{bmatrix} = [0.00015]$$

An analysis of variance table also requires calculation of the sum of squares corrected for the mean (matrix T^tT) and the factor or regression sum of squares (matrix F^tF) which can be obtained from the responses:

$$T = \begin{bmatrix} y_1 \\ y_2 \\ y_3 \end{bmatrix} - \begin{bmatrix} \bar{y} \\ \bar{y} \\ \bar{y} \end{bmatrix} = \begin{bmatrix} 0.30 \\ 0.46 \\ 0.59 \end{bmatrix} - \begin{bmatrix} 0.45 \\ 0.45 \\ 0.45 \end{bmatrix} = \begin{bmatrix} -0.15 \\ 0.01 \\ 0.14 \end{bmatrix}$$

where

$$T = Y - \bar{Y}$$

Therefore,

$$T^tT = \begin{bmatrix} -0.15 & 0.01 & 0.14 \end{bmatrix} \begin{bmatrix} -0.15 \\ 0.01 \\ 0.14 \end{bmatrix} = [0.0422]$$

The simplest way of obtaining the *TreatmentSS* is to subtract the *ResidSS* from *TotalSS*:

$$TreatmentSS = 0.0422 - 0.00015 = 0.04205$$

The analysis of variance table can now be constructed:

Source of variation	d.f.	SS	MS	Variance ratio
Total (corrected)	1	0.0422		
Treatment	1	0.04205	0.04205	280.33
Residual	1	0.00015	0.00015	

Clearly most of the variance in the data is accounted for by the regression and the not-unexpected high variance ratio of 280.33 exceeds the critical F value of 161.3 for one and one *d.f.* at a probability level of $P = 0.05$. The model would therefore seem to provide an adequate fit to the data.

****

SAQ 5.4a	Determine the least-squares estimates for the intercept (b_0), slope (b_1) and curvature (b_{11}) in the single-factor example given its $X'X$ matrix made up from the matrix of parameter coefficients X, containing a column of 1's for the intercept, a column containing the values 1, 2, 3 and 4 for the levels used, and a column containing the values 1, 4, 9 and 16 for the squares of the coefficients.

$$X'X = \begin{bmatrix} 1 & 1 & 1 & 1 \\ 1 & 2 & 3 & 4 \\ 1 & 4 & 9 & 16 \end{bmatrix} \begin{bmatrix} 1 & 1 & 1 \\ 1 & 2 & 4 \\ 1 & 3 & 9 \\ 1 & 4 & 16 \end{bmatrix}$$

$$= \begin{bmatrix} 4 & 10 & 30 \\ 10 & 31 & 100 \\ 30 & 100 & 354 \end{bmatrix}$$

Response

My estimates of the parameters are given in the vector $\widehat{B}$.

$$\widehat{B} = \begin{bmatrix} 3.5 \\ 5.9 \\ -0.5 \end{bmatrix} \begin{matrix} = b_0 \\ = b_1 \\ = b_{11} \end{matrix}$$

I again estimated these parameters by minimizing the squares of the residuals using the generalized least-squares matrix equation (5.4). This required calculation of the inverse of $X'X$ by the method given previously which is

$$\left(X'X\right)^{-1} = \begin{bmatrix} 7.75 & -6.75 & 1.25 \\ -6.75 & 6.45 & -1.25 \\ 1.25 & -1.25 & 0.25 \end{bmatrix}$$

and

$$X^tY = \begin{bmatrix} 68.0 \\ 192.0 \\ 618.0 \end{bmatrix}$$

Therefore,

$$\hat{B} = \begin{bmatrix} 7.75 & -6.75 & 1.25 \\ -6.75 & 6.45 & -1.25 \\ 1.25 & -1.25 & 0.25 \end{bmatrix} \begin{bmatrix} 68.0 \\ 192.0 \\ 618.0 \end{bmatrix} = \begin{bmatrix} 3.5 \\ 6.9 \\ -0.5 \end{bmatrix} \begin{matrix} = b_0 \\ = b_1 \\ = b_2 \end{matrix}$$

$$\hat{B} = \qquad (X^tX)^{-1} \qquad\qquad X^tY$$

These parameters may now be plotted as graph of response versus reactant concentration.

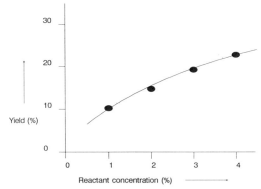

Fig.5.4a. *Plot of the response (% yield) against reactant concentration with the line for the second-order equation fitted using the calculated parameters*

You will have already estimated parameters using the generalized least-squares equation and this particular one should not have been too difficult if you know how to calculate the inverses of 3×3 matrices. If you found you couldn't do this re-read the part of Section 5.4 on the inversion of 3×3 matrices. Once you had the parameter estimates, you should have calculated the predicted responses using the equation:

$$\hat{Y} = \qquad X \qquad\qquad \hat{B}$$

$$\hat{Y} = \begin{bmatrix} 1 & 1 & 1 \\ 1 & 2 & 4 \\ 1 & 3 & 9 \\ 1 & 4 & 16 \end{bmatrix} \begin{bmatrix} 3.5 \\ 6.9 \\ -0.5 \end{bmatrix} = \begin{bmatrix} (1 \times 3.5 + 1 \times 6.9 + 1 \times -0.5) \\ (1 \times 3.5 + 2 \times 6.9 + 4 \times -0.5) \\ (1 \times 3.5 + 3 \times 6.9 + 9 \times -0.5) \\ (1 \times 3.5 + 4 \times 6.9 + 16 \times -0.5) \end{bmatrix} = \begin{bmatrix} 9.9 \\ 15.3 \\ 19.7 \\ 23.1 \end{bmatrix}$$

and the differences between these and the actual responses determined (R). Therefore,

$$R = \begin{bmatrix} 10 \\ 15 \\ 20 \\ 23 \end{bmatrix} - \begin{bmatrix} 9.9 \\ 15.3 \\ 19.7 \\ 23.1 \end{bmatrix} = \begin{bmatrix} 0.1 \\ -0.3 \\ 0.3 \\ 0.1 \end{bmatrix}$$

The sum of the squares of the residuals is given by R^tR:

$$R^tR = [0.1 \quad -0.3 \quad 0.3 \quad 0.1] \begin{bmatrix} 0.1 \\ -0.3 \\ 0.3 \\ 0.1 \end{bmatrix} = [0.2]$$

Having worked through most of the matrix manipulations several times by now, complete matrix derivations of the quantities for solution of the generalized solution of linear equations by least-squares techniques will be skipped over from now on unless something new is introduced. Therefore, the other quantities required for the example analysis of variance have been worked out as previously and the *ANOVA* table is presented below.

Source of variation	d.f.	SS	MS	Variance ratio
Total (corrected)	3	98		
Treatment	2	97.8	48.90	244.50*
Residual	1	0.2	0.2	

* exceeds critical F value for 1, 2 d.f. at $P = 0.05$ probability level.

****

SAQ 5.4b   For the single-factor example, calculate the coefficient of multiple correlation (R^2) and correlation coefficient (r), and compare these with the F ratios calculated previously. Does the second-order model provide a better fit than the first-order model?

Response

For the first-order model R or $r = 0.9939$ and for the second-order model $R = 0.9989$, a small increase which seem to indicate a better fit. However, in Section 5.3.3 and in response to SAQ 5.4a the values of F were calculated to be 161.33 for the first-order model and 244.50 for the second-order model respectively. Both these F values exceed the appropriate critical F values at the $P = 0.05$ probability level, but only the first-order model is significant at the $P = 0.01$ level. The calculated F ratio of 161.33 for the first-order model is greater than the tabulated F value for one and two *d.f.* at $P = 0.01$ (98.5), whereas the second-order F ratio is less than the tabulated F for two and one *d.f.* at $P = 0.01$ (4999). Therefore, in spite of an increase in the value of R, if I was using a probability level of $P = 0.01$, I could not accept the second-order model and would have to accept the first-order model.

****

SAQ 5.4c	Consider two designs based on the model $y_i = \beta_0 + \beta_1 x_{1i} + r_i$ in which only two levels are used, the first level being the same for both designs and the second level being different, such that the X matrices are as given below.

Design 1

$$X_1 = \begin{bmatrix} 1 & 1 \\ 1 & 3 \end{bmatrix}$$

Design 2

$$X_2 = \begin{bmatrix} 1 & 1 \\ 1 & 10 \end{bmatrix}$$

What effect does changing the level of the second run have upon the estimates of uncertainty in the parameters? How does this influence the choice of experimental design if the experimenter wants as small as possible an estimate of uncertainty for b_1?

Response

In order to compare the effects of the designs, the inverse matrix $(X'X)^{-1}$ has to be calculated in each case. Each $X'X$ matrix is found by pre-multiplying the X matrix by the transformed matrix, X', as given below:

Design 1

$$X'X_1 = \begin{bmatrix} 2 & 4 \\ 4 & 10 \end{bmatrix}$$

Design 2

$$X'X_2 = \begin{bmatrix} 2 & 11 \\ 11 & 101 \end{bmatrix}$$

The determinants are:

$$D_1 = 2 \times 10 - 4 \times 4 = 4 \qquad D_2 = 2 \times 101 - 11 \times 11 = 81$$

Having used methods previously described, the $(X'X)^{-1}$ matrices were calculated to be

$$(X'X)_1{}^{-1} = \begin{bmatrix} 2.5 & -1 \\ -1 & 0.5 \end{bmatrix} \qquad (X'X)_2{}^{-1} = \begin{bmatrix} 1.25 & -0.135 \\ -0.135 & 0.025 \end{bmatrix}$$

Increasing the spacing between the levels used has decreased the values of all the elements in the $X^t X^{-1}$ matrices. The estimates of uncertainty in the parameters would therefore be smaller in the second design. However, a closer examination would seem to show that the uncertainty in b_0 would be halved if the same responses were obtained, whereas the uncertainty in b_1 would be much smaller, with the value of the bottom right-hand element decreasing by a factor of 20. You can then be more confident in the estimate of b_1 for the second design. In addition, the estimates of uncertainty in the covariances are also smaller in the second design. The significance of this will be dealt with in the next sub-section. In conclusion, if you wish to be as confident as possible about an estimate of b_1, you should space the levels as widely as is practically possible.

$$**********************************$$

SAQ 5.7a
> What is the axial spacing required for a two-factor central-composite rotatable design with eight centre points? Is this design orthogonal?

Response

Using Eq. 5.20 you can determine the axial spacing required for an orthogonal design. Thus

$$a^2 = \frac{\sqrt{(N_c + N_a + N_o)N_c} - N_c}{2} = \frac{\sqrt{(4 + 4 + 8)4} - 4}{2}$$

$$a^2 = (\sqrt{64 - 4})/2 = 2$$

and $a = \sqrt{2} = 1.414$. For the design to be rotatable the relationship represented in the equation

$$a^4 = N_c$$

has to be true. Since $N_c = 4$, $a^2 = \sqrt{4} = 2$ and $a = 1.414$. Therefore, the values for axial spacing in the design to be both orthogonal and rotatable coincide.

****

SAQ 5.7b	What are the co-ordinates of the stationary point on the response surface for the equation given below?
	$$y_i = 5 + 2x_{1i} + 2x_{2i} + 3x_{1i}^2 + 3x_{2i}^2 - 4x_{1i}x_{2i}$$
	What is the response at this stationary point? What shape will the response surface take?

Response

The calculations required for finding the co-ordinates of the stationary point and its response are as follows. Firstly, it is necessary to define a matrix S consisting only of second-order terms from the calculated second-order equation such that

$$S = \begin{bmatrix} b_{11} & b_{12}/2 \\ b_{12}/2 & b_{22} \end{bmatrix} = \begin{bmatrix} 3 & -2 \\ -2 & 3 \end{bmatrix}$$

A matrix f containing the first-order parameters b_1 and b_2 is also required:

$$F = \begin{bmatrix} b_1 \\ b_2 \end{bmatrix} = \begin{bmatrix} 2 \\ 2 \end{bmatrix}$$

Using Eq. (5.27) it is possible to define the co-ordinates of the stationary point on the response surface.

$$s = -0.5S^{-1}f$$

The stationary point co-ordinates may now be calculated:

$$S = \begin{bmatrix} x_{1s} \\ x_{2s} \end{bmatrix} = 0.5 \begin{bmatrix} 3 & -2 \\ -2 & 3 \end{bmatrix}^{-1} \begin{bmatrix} 2 \\ 2 \end{bmatrix}$$

The determinant of S is $(3 \times 3) - (-2 \times -2) = 5$. The inverse matrix S^{-1} is

$$S^{-1} = \begin{bmatrix} 3/5 & 2/5 \\ 2/5 & 3/5 \end{bmatrix} = \begin{bmatrix} 0.6 & 0.4 \\ 0.4 & 0.6 \end{bmatrix}$$

The stationary point co-ordinates are then given by

$$S = \begin{bmatrix} x_{1s} \\ x_{2s} \end{bmatrix} = -0.5 \begin{bmatrix} 0.6 & 0.4 \\ 0.4 & 0.6 \end{bmatrix} \begin{bmatrix} 2 \\ 2 \end{bmatrix} = \begin{bmatrix} -1.0 \\ -1.0 \end{bmatrix}$$

The co-ordinates of the stationary point are $x_{1s} = -1.0$ and $x_{2s} = -1.0$. The response at these co-ordinates is given by entering these into the equation describing the response surface:

$$y_s = 5 + (2 \times -1.0) + (2 \times -1.0) + (3 \times -1.0^2) + (3 \times -1.0^2) - (4 \times -1.0 \times -1.0)$$

$$= 5 - 2.0 - 2.0 + 3.0 + 3.0 - 4.0 = 3.0$$

The response at this stationary point is 3.0. This is less than the intercept term (b_0) which is the response at co-ordinates 0,0 for factors x_1 and x_2, and the pure second-order terms b_{11} and b_{22} are positive and have the same magnitude. This indicates that the response surface is a parabolic bowl opening upwards (like an upside-down umbrella).

****

Index